Effective Technical Communication: A Guide for Scientists and Engineers

Barun K. Mitra
Formerly Professor of English,
Indian Institute of Technology Kharagpur

OXFORD

UNIVERSITY PRESS

OXFORD
UNIVERSITY PRESS

YMCA Library Building, Jai Singh Road, New Delhi 110001

Oxford University Press is a department of the University of Oxford.
It furthers the University's objective of excellence in research, scholarship,
and education by publishing worldwide in

Oxford   New York
Auckland   Cape Town   Dar es Salaam   Hong Kong   Karachi
Kuala Lumpur   Madrid   Melbourne   Mexico City   Nairobi
New Delhi   Shanghai   Taipei   Toronto

With offices in
Argentina   Austria   Brazil   Chile   Czech Republic   France   Greece
Guatemala   Hungary   Italy   Japan   Poland   Portugal   Singapore
South Korea   Switzerland   Thailand   Turkey   Ukraine   Vietnam

Oxford is a registered trade mark of Oxford University Press
in the UK and in certain other countries.

Published in India
by Oxford University Press

ISBN-13: 978-0-19-568291-5
ISBN-10: 0-19-568291-2

Typeset in Garamond
by Innovative Processors, New Delhi
Printed in India by Radha Press, Delhi 110031
and published by Manzar Khan, Oxford University Press
YMCA Library Building, Jai Singh Road, New Delhi 110001

In Memoriam

*My beloved wife Sutapa—The soul behind all
my endeavours for nearly fifty years*

Preface

SCIENTISTS, engineers, and researchers often use the medium of the English language to communicate, be it for research papers, sharing their research findings with peers, technical reports, or speeches. To be effective, it is imperative that all authors are well-conversant with all aspects of various communication techniques. This can enable them to have a clear idea of the ways through which one can express one's thoughts and ideas whenever required.

In this age of science and technology, the communication process has acquired an added criticality where everyday communication within and among groups is essential. Individuals have to be in constant touch with one another by means of diverse types of communication processes to realize a common goal. There cannot be any cohesion amongst the members of a group without proper interaction. Therefore, it is essential that there is a smooth flow of communication not only within peer groups, but also up and down the hierarchy. In the corporate sector, for example, cooperation and interaction involving a number of individuals are essential.

Communication that flows from the higher echelons of a particular corporate sector, that is, from the CEO or Managing Director of a company to those who are in the next levels of the organizational ladder, requires the downward flow of communication. The upward flow of communication is exactly the opposite. Here communication emanates from somebody in the lower levels and flows to individuals higher up in the management hierarchy. Finally, peer communication flows horizontally amongst those who are at the same level of hierarchy. All these flows are possible only with communication which is clear, precise, logical, and easy to follow.

This book is the fruit of my experience in teaching English for three decades to budding scientists and technologists at the Indian Institute of Technology (IIT) Kharagpur. Also, as a Professor of English at IIT, I coordinated some courses sponsored by the Indian Society for Technical Education (ISTE) and Quality Improvement Programme (QIP). These courses, on 'English for Scientists and Technologists' and on 'Communication Skills', were focused on teachers of engineering colleges. Subsequently, as Director and Vice-President of the South Asia Hankes Foundation, while working on a language project at Minneapolis and later in India, I got the fillip for writing such a book based on of the work I had been doing and the ideas emanating therefrom.

I believe from my experience that this book will be of immense help to scientists and engineers engaged in various industries—both national and multinational. It will also be of help to persons engaged as faculty members in reputed colleges and universities. This book will also benefit research students writing technical papers as part of their research curricula.

This book discusses various communication techniques that are extremely relevant for writing technical reports. It also covers the various types of technical reports that can be written. Stress has been laid on clarity, simplicity, style, and brevity. This book also discusses the principles of sentence formation and paragraph construction. The formats of technical reports have been discussed along with the proper way of communicating via e-mail. The suitable uses of charts, graphs, and audio-visual aids have also been detailed.

In addition, the book discusses the fundamentals of a successful speech. The processes of speech organization and public speaking have also been examined. Various suggestions have been made to help authors prepare and deliver their speeches. A later chapter also looks at the evolution, use, and appropriateness of gender neutral language.

This book includes a discussion on how the clarity of technical texts can be assessed by using the fog index or clarity index. An acquaintance with this method would help scientists and engineers to ascertain the clarity of their writing. Finally, this book lists the pitfalls that one may come across when writing and provides suggestions on how to avoid them.

This book, thus, caters to the essential requirements of scientists, engineers, and researchers, and provides the readers with a logical sequence which, if followed carefully, can be of great help in their technical communication. A thorough understanding of the techniques illustrated in this book will help the readers pursue their vocations efficiently, confidently, and successfully, and will act as a zealous guide in their various professional activities.

Barun K. Mitra

Acknowledgements

Iowe thanks to many people who helped me develop my thoughts on this subject. The credit for persuading me to write this book should go to my son Biswadip, himself an Electronics and Computer Engineer. His gentle persuasion coupled with his earnest endeavour to help me in every way, especially in technical matters, made this possible. I would like to thank Prof. S. S Bhattacharyya, Dept. of Material Science (Atomic and Molecular Section), Indian Association for the Cultivation of Science (Kolkata), for helping me with a number of scientific papers published in reputed journals. I also appreciate the discussions on this subject that I have had with Professor Saswati Lahiri, Dept. of Organic Chemistry, Indian Association for the Cultivation of Science, Kolkata.

In addition, I wish to thank my erstwhile colleagues at IIT Kharagpur, who enriched my thought process through the various discussions that we had. I sincerely respect their deep insight in the area of technical communication. I would also like to thank many past students of IIT Kharagpur who have been inadvertent motivators of this book. I thank my daughter-in-law, Mrs. Basundhara Mitra, who helped me in preparing the typescript.

Another person whose encouragement and constant motivation helped me complete the work is my daughter Gopa (Dr Swagata Ghosh). Thank you Gopa. Finally, I would like to thank the editorial staff at Oxford University Press, for their feedback and suggestions at every stage of this book.

Barun K. Mitra

Table of Contents

Introduction

THE rapid industrialization which began in the 20th century had a great impact on communication. Old methods of communication gave way to faster and more effective ones. While it is not certain when humans first began to communicate, hieroglyphs or words written on papyrus leaves have been in existence for thousands of years. Gradually, communication through the medium of paper became a part of cultures across the world. It retains its historic popularity till today. However, language is dynamic and evolves continually. The development of modern technology fuelled the need for a specialized form of English to cater to its distinct communication needs.

English for scientists and engineers gradually evolved to make technical developments intelligible to the scientific community as well as the wider public. It included carefully chosen words or scientific vocabulary and thoughtfully constructed sentences and paragraphs. This book aims to make scientific and engineering writers conscious of these evolving communication techniques. It will also help them to fine tune their writing so that they can communicate their ideas effectively, at the same time observing scientific decorum.

Scientific and engineering writing presents information clearly and concisely without ambiguity, empty phrases, and repetition. There is no place for rambling or unnecessary embellishment. Attention to grammar and style is required. Every word has to be chosen carefully to present information effectively. Scientists and engineers, in fact, should be conscious of the fact that they have a very useful tool—the English language—in their hands, which, if used well, will enable them to reach out to a wider audience.

Language is systematic communication through vocal symbols. Each language is essentially a code, that is, a group of symbols that can be structured to be meaningful to others. The English language, like any other language, is a code. It contains words that are arranged in a meaningful grammatical structure or order. However, scientists must make certain that this encoding is done in such a way that the recipient is able to decipher the code easily.

Communication does not end with encoding or decoding a message. The word 'communicate' in Latin means 'exchange information'. The main objective of communication is to facilitate the exchange of thoughts and ideas in an effective and methodical way.

There are different methods and techniques in the communication process for different requirements. A scientist or an engineer, in professional life, should be well-acquainted with the techniques needed for different types of communication, such as writing a scientific or technical paper for a reputed journal, delivering a speech, writing a technical report, or simply writing an e-mail.

Before writing on any technical subject, a scientist or an engineer must have a clear idea of the principles of scientific vocabulary and must be able to choose the right words and discard ambiguities, haziness, long words, etc. This is discussed in the opening chapter. The next chapter discusses the principles of framing sentences and the technique of paragraph construction.

The following chapters in the book, Writing Scientific and Engineering Papers, Technical Guidelines for Communication, and Effective Use of Charts, Graphs, and Tables, provide information about the kind of communication required for scientific and engineering journals. Scientists or engineers, whatever might be the chosen areas of their work, have to communicate their findings to others. They may need to have a paper published in a reputed journal. This task is not easy. A well-written scientific paper must adhere to certain technicalities, including a specified format, which cannot be ignored.

Scientists and engineers are aware of the benefits that they can derive if their paper gets published in a well-known scientific journal. Not only is it important for them to be the first to publish a discovery or a new finding, but also to have their work published in a journal where it will receive the widest attention. According to a newspaper report, the first scientific journal was published in France in 1665. Since then, the number of scientific journals has grown enormously. According to an estimate on the website of *The Hindu* (8/11/2003) more than a million scientific papers are published annually by over 20,000 journals.

There are other facets of scientific and technical writing that are just as important. Diverse topics of importance to scientists or engineers have been dealt with in the chapters: A Scientific Approach to Writing E-mail, Speech Communication: Effective Presentation Techniques, and Writing Technical and Business Reports. A chapter has also been devoted to another important aspect: Observing the Code of Gender Neutral Language.

The book ends with a note of caution. Whatever the subject of the scientist or engineer's writing, the foundation of the entire structure stands on correct and flawless English. So even those who have a good command over the English language should be aware of the pitfalls that may beguile them. The final chapter: Look Before you Write: Avoiding the Pitfalls of Common Grammatical and Spelling Errors, flags potential oversights and mistakes.

All these chapters focus on one objective. This is to make scientists and engineers fully conversant with the key communication techniques necessary for different genres of writing, which can be of immense help to them in their profession.

1
Principles of Scientific Vocabulary

All big things have little names,
Such as life and death, peace and war,
Or dawn, day, night, love, home.
Learn to use little words in a big way.

—Anonymous

SCIENCE (from the Latin word 'scientia' meaning knowledge) is knowledge arranged methodically or systematized knowledge. Its progress is marked by the emergence of the scientific method, which rests on the rational, accurate, and clear exposition of facts. At its core is a thoroughness of approach that is needed to establish any new finding. For this, the use of appropriate words and sentences is essential.

A sentence is a combination of words, which are the basic units of expression. To communicate effectively with others and to disseminate ideas, views, and observations to a broader community, a scientist or an engineer needs a thorough knowledge of the intricacies of the language. The entire structure of scientific or engineering communication stands on the solid foundation of words. Hence, superficial knowledge of a word and its spelling is of little help. The writer must choose his/her words carefully so as to convey his/her ideas most effectively.

The ability to choose the appropriate words and use these in the appropriate context comes from familiarity with words and their usage. Thus, a writer should avoid using words with which he/she is unfamiliar.

There are eight principles for choosing appropriate words and phrases in scientific and engineering communication. These have been listed below.

- Use short and simple words.
- Use compact substitutes for wordy phrases.
- Avoid redundant words and expressions.
- Avoid the use of mixed metaphors and other figures of speech.
- Avoid hackneyed and stilted phrases.
- Avoid verbosity in the use of common prepositions.
- Avoid the incorrect use of words.
- Exercise care while using technical terms borrowed from traditional English.

Some examples of each of these principles have been given below.

1.1 Use Short and Simple Words

Any communication by scientists and engineers requires the use of short and simple words, which are more forceful than long words. A short word appropriately used, enhances the clarity of expression.

Examples of a few short and simple words that should be used instead of polysyllabic words are provided below:

Avoidable words	Recommended words
accomplish	do
terminate	end
conflagration	fire
adjacent to	near
cognizant of	aware of
perform	do
vehicle	car

attempt	try
beverage	drink
envisage	foresee
ameliorate	improve
endeavour	try
perchance	perhaps
abhorrence	hatred
viable	workable
deem	think

1.2   Use Compact Substitutes for Wordy Phrases

The use of simple substitutes greatly improves the clarity of written communication. Some examples are given below.

Avoidable usage	*Recommended usage*
Take into consideration	consider
Avail yourself of	use
In view of the fact that	since
In all instances	always
In a very small number of cases	rarely
In view of the fact that	because
Subsequent to	after
In the near future	soon
In spite of the fact that	although
As a consequence of this fact	consequently
A small number of	a few
Despite the fact that	although
At the present time	now
Prior to	before
For the duration of the study	during the study
During the process of	during
Checked for the presence of	checked for
A series of observations	observations
In order to provide a basis for comparing	to compare

Make an adjustment in	adjust
Give encouragement to	encourage
Is equipped with	has

1.3 Avoid Redundant Words and Expressions

The clarity of the text can be improved by completely removing expressions that do not add value to the text. Every sentence should be trimmed down to its essentials. A number of redundancies can creep into a written communication if it is written carelessly.

Examples of expressions that should be avoided have been given below.

- It has been found that
- It is interesting to note that
- As already stated
- It may be said that
- It is worth mentioning at this point

A list of some expressions, parts of which should be removed due to their redundancy, has been given below. The redundant words have been placed in brackets.

- (absolutely) essential
- (advance) planning
- (advance) warning
- (as) for example
- (at) about, at (about): use any one
- at (the) present (time)
- (brief) moment
- during (the course of)
- merged (together)
- reply (back)
- is (now) pending
- eradicate (completely)
- (current) trend
- never (before)

- (new) innovation
- (mutual) cooperation
- (close) proximity
- (necessary) requisite
- (protrude) out
- revert (back)
- (free) gift

There are a number of instances when the verb is capable of expressing the notion of togetherness. In such instances, the expression 'together' is superfluous. The examples below illustrate this point.

- meet (together)
- unite (together)
- connect (together)
- join (together)
- mix (together)

A little thinking can help a writer avoid such redundancies. Some of these redundancies can be used in special cases for emphasis. However, it is for the scientist or engineer to judge the appropriate context and not use them indiscriminately.

An example of how redundancies can lead to heaviness in the text is:

The main cause for the failure of the experiment was that adequate precaution was not taken at the time of the experiment for preventing such kind of accidents.

The sentence can be expressed more simply by using fewer words:

The experiment failed, as adequate precautionary measures were not taken.

Here is another example:

If the supply of drawing papers already sent falls short of your demands, application has to be made to the Stores Officer for further supply.

This could be shortened to:

If more drawing paper is needed, apply to the Stores Officer.

In the example below, a 34-word sentence is shortened to an 11-word sentence without any change in the meaning:

At the present moment our administrative unit has already initiated the procedure of inviting applications from those who are in a position to offer themselves as candidates for the post of Senior Mechanical Engineer.

This could instead be written as:

Applications have been invited for the post of Senior Mechanical Engineer.

1.4 Avoid the Use of Mixed Metaphors and other Figures of Speech

Sometimes, the use of metaphors can lead to lack of clarity in a text. Usually, the use of metaphors in scientific or engineering text is bad. The use of mixed metaphors is worse.

For example, the following sentence uses mixed metaphors:

Instead of beating about the bush put your cards on the table.

It uses two different metaphors: 'beating about the bush' and 'putting one's cards on the table'.

The sentence can instead be written simply as:

Clearly say what you have to state.

There is a general notion that one should avoid words that may be considered vulgar. The origin of this notion may be traced back to the Victorian era when prudishness was at its peak. (A prude is a person who is excessively modest in his/her behaviour, dress, and speech). This has led to the use of euphemisms, that is, words that are less direct. These words are not distasteful and are never vulgar.

Some examples are listed here.

- In the family way: This should be preferred over 'pregnant' (as used in HR policies for industries, for example)
- Sales representatives: This should be preferred over 'salesmen' (as used in titles used in industries, for example)
- Lower income group: This should be preferred over 'the poor' (as used in certain segments addressed by a firm, for example)

Of course, expressions like 'sales representatives' or 'lower income group' are neutral expressions and should, therefore, be given preference.

1.5 Avoid Hackneyed and Stilted Phrases

Engineers and scientists must try to overcome the temptation to use over-worn phrases and expressions. These expressions are used by greenhorns (first-time writers) to embellish their language. They use these phrases in a naive attempt to show their knowledge.

Here are some common examples of such hackneyed expressions:

- filthy lucre
- olive branch
- strain every nerve
- flying colours

1.6 Avoid Verbosity in the Use of Common Prepositions

There is no dearth of appropriate prepositions to suit any context in the English language. However, the tendency amongst writers is to use imposing expressions instead of simple prepositions. This tendency can lead to an avoidable lack of clarity in communication.

Common examples of these expressions are:

As to: In the sentence, 'The director should give a clear indication *as to* the policies he wants to introduce', the simple preposition 'of' could serve the purpose just as well.

In relation to: In the sentence, 'The salaries of the scientists vary *in relation to* their qualifications and experience', the preposition 'with' provides the same meaning.

With regard to: In the sentence, 'The recruitment of only two extra scientists will make little difference *with regard to* the output of work', the preposition 'to' will convey the same meaning.

Since verbose writing obscures the meaning of a sentence and only taxes the readers' time and patience, it should be avoided.

1.7 Avoid the Incorrect Use of Words

Scientists and engineers need to be aware of the specific meanings of the words they use. They have to choose words that can accurately and precisely express their ideas. If a writer uses a word she/he is not absolutely certain about, it can end up creating confusion in the minds of the readers.

Using words accurately is extremely important for scientists and engineers. There are words that are synonymous and are more or less similar in meaning. However, only a writer who is conversant with the different meanings that may be attributed to a word can choose the correct word for the correct context.

Here are some words that describe 'something that happens'. There is a slight difference in the meaning of each word.

Happening: It refers to some unusual or strange incident that has happened.

For example: The employees were afraid of the strange happenings in the vicinity of their company premises.

Incident: It refers to a course of action that may not be very important, yet has the air of something unexpected or unusual that makes one remember it.

For example: I can recall the incident that led to the employee being warned by his supervisor.

Event: It describes a circumstance that is important, because of which it is remembered.

For example: The invention of the integrated circuit was a major event in the history of the electronics industry.

Occurrence: This is a formal word that indicates a happening that is either common or rare.

For example: These days, having an open office is a very common occurrence.

For example: Robbery is a rare occurrence in this small town.

Each of these words has a unique meaning. In scientific or engineering communication, it is important not to use such words interchangeably. Likewise, the word 'keep' is not the same as 'put', as shown in the following example:

The teacher *keeps* the book on the shelf.

The teacher enters the classroom and *puts* the book on the table.

Similarly, 'humid', 'damp', and 'moist', or 'pliable' and 'flexible', or 'permeate' and 'percolate', do *not* express the same meanings.

There are some words that have become so popular that writers use them indiscriminately, ignoring their precise meaning. The word 'blueprint', for example, has become an attractive alternative to 'scheme' or 'plan'. But this term, which comes from engineering technology, actually stands for the *final* stage of paper design.

There are many instances in which ponderous words are misused only because the writer does not know its meaning. Some examples are:

Syndrome: This term, which is gaining popularity, can be easily misused by a writer who is not certain of its meaning. The word 'syndrome' is a medical term. It actually means a group of symptoms, which collectively suggest a particular disease. The syndrome itself is not a disease.

Synergism: The precise meaning of this word is 'the simultaneous collective action which an effect greater than the sum of the individual effects'. It does not mean just collective action, which it is erroneously used to refer to sometimes.

Catalyst: This is a very popular term, which should be used with discretion to avoid incorrect usage. Catalysis is the speeding up of the role of chemical reaction by the addition of some substance, which undergoes no chemical change itself. A catalyst is the agent that brings about this change. This context, which includes 'no change to itself', should be borne in mind before using this word.

There are many other such words, such as conductivity, frequency, anodise, pneumatic, alignment, backlash, impedance, permeability, servo-mechanism, etc., that should be used in proper perspective. The language of science and engineering is never static; it is dynamic. Hence, the scientist or engineer must keep track of these words and their *current* meaning prior to using them.

1.8 Exercise Care while using Technical Terms Borrowed from Traditional English

Common English words are used increasingly to express various scientific and engineering matters. A majority of these words are used in Computer Science and Engineering. A scientist or an engineer writing a document must ensure that these scientific or engineering terms are used appropriately and not confused with their traditional meanings.

Some examples of scientific or engineering terms that have roots in traditional English, have been provided below.

Heap	A temporary data storage area where random access is possible.
Hierarchy	It denotes the method in which data is organized in a step-wise order.
Howler	A buzzer that helps the telephone exchange operator detect whether the telephone user's handset is on the receiver.

Junction box	An electrical unit where it is possible to get a number of electrical wires connected together.
Packet	A group of data bits that can be transmitted together as a group.
Specific	The electrical charge of an elementary particle divided by its mass.
Warm up	When a machine, after being switched on, is in the process of reaching its optimum state.
Idle	A particular state of the engine in which, though running, it does not provide power to move any vehicle or aircraft.
Hang	The particular state when a computer is held in an endless loop and fails to respond.
Footprint	The area covered by any transmitting device such as an antenna or a satellite.
Jump	A term in computer parlance when a programming command is given to direct the processor to a different section of the programme.
Declare	A term used to define a computer programme variable.
Half-life	The time taken for half the atoms in a radioactive isotope to decay.
Inductive	The production of electrical current in a conductor by a change of magnetic field.
Thread	A programme in a computer consisting of many independent smaller sections or heads.
Mouse	A small hand-held input device in a computer. It is used to control the position of the cursor on the computer screen.

To conclude, English written by engineers and scientists must be simple and precise. Long, abstract, fancy, and redundant phrases must be avoided. Short and simple words build the structure, followed by sentences, paragraphs, and ultimately the whole text of the communication. Hence, the choice of words is of great importance. It has to be kept in mind that scientific or engineering writing has a definite objective. Its main purpose is to communicate something. Hence, the words and sentences that are used should be simple, clear, brief, and unambiguous. Without these attributes, communication of any sort is bound to become boring, stilted, and foggy. Words, if used appropriately, go a long way in building the structure of sentences and paragraphs.

A quote from H.W. Fowler reveals the basic rules to be followed in the case of written communication:

Prefer the familiar word to the far-fetched.
Prefer the concrete word to the abstract.
Prefer the single word to the circumlocution.
Prefer the short word to the long.
Prefer the Saxon word to the Romance.

2

Techniques of Sentence and Paragraph Construction

A sentence—long or short
always says something
clearly, unequivocally,
and leads on to the paragraph

THE concise expression of an idea is fundamental to scientific and engineering communication. Instead of using clumsy expressions and unnecessary words, ideas should be expressed as precisely as possible. Brevity is most important.

In this context, we will discuss the following three aspects:

1. Principles of framing a sentence in scientific and engineering communication,
2. techniques of paragraph writing, and
3. measuring the clarity of a text through fog index or clarity index.

2.1 Principles of Framing a Sentence

A sentence is a unit of thought. It is a collection of words syntactically and grammatically arranged to convey a certain meaning. The main objective of scientific or engineering text is to convey views, observations, and findings to the scientific or engineering community or society at large. Hence, words should be arranged in a sentence in such a way that the reader gets a clear idea of what the writer wants to convey.

A paragraph usually starts with a *topic sentence* that expresses the main idea of the writer. This topic sentence also provides the reader of the paper an idea about the main theme of the paragraph, which will be gradually developed in subsequent sentences. As language is a code, the sentence should be framed in such a well-ordered way that the reader finds it easy to decode (understand) it. To do this, the writer must adhere to certain rules while framing a sentence.

The Ten Principles of Framing a Sentence

1. Avoid long sentences. Do not use unnecessary words.

   For example, the sentence below has several redundant words.

   Original sentence: There is a cell called the photo-electric cell, which is characterized by its ability to change its electrical characteristics when light falls upon it.

   Revised sentence: Photo-electric cells change their electrical characteristics when exposed to light.

2. Do not cram different points into one sentence. Each sentence should deal with one point clearly.

   For example, two very distinct observations are being made in the following sentence. It should be revised so that each observation is made clearly in a different sentence.

   Original sentence: The experiment gives a clear indication of the effect of lower voltage on the power dissipation, while also showing a strong correlation between high performance and excessive standby current.

   Revised sentence: The experiment gives a clear indication of the effect of lower voltage on the power dissipation. It also shows a strong correlation between high performance and excessive standby current.

3. Do not twist the order of the words. Words should be arranged in a logical sequence. After-thoughts should not be added to a sentence.

   In the example below, the comment on the components of the product precedes that on the full product.

   Original sentence: The entire product development was completed in just six months and each component of the product was imported in a ready-to-use state.

   Revised sentence: Each component was imported in a ready-to-use state, leading to the entire product development taking only six months.

4. Ensure that similar ideas that are expressed in a sentence should take the same structural and grammatical form.

   This is illustrated in the example below. Here, the original sentence has elements (processing, measuring, and analysis), which are not grammatically consistent and need to be rewritten.

   Original sentence: The processing, measuring, and analysis of the solution need to be done diligently.

   Revised sentence: Processing, measuring, and analyzing the solution need to be done diligently.

5. Make positive statements without being hesitant or non-committal.

   This is illustrated in the example below. Here, the original sentence is very long and can be revised to make it direct.

   Original sentence: I do not think that the way the work progressed was satisfactory.

   Revised sentence: The progress was unsatisfactory.

6. Do not use pompous words and phrases.

   This is illustrated in the example stated here. The expression 'spurned with disdain' can be effectively replaced with the word 'rejected' and 'made to him' can be removed due to redundancy.

Original sentence: He spurned with disdain the offer made to him.

Revised sentence: He rejected the offer.

7. Use active instead of passive voice. A sentence in active voice is direct and is easier to read and understand. Moreover, a sentence becomes forceful when it is in active voice. However, when the action is more important than the person performing the action, it is better to use passive voice, as in the following example:

Original sentence: The doctor controlled the growth of cancer cells through chemotherapy.

Revised sentence: The growth of cancer cells was controlled by chemotherapy.

8. Use the word 'respectively' carefully. It often makes the meaning of a sentence rather hazy. The word actually means 'in regard to each in the order named'. For example, the original sentence below is confusing, while the revised one is clear.

Original sentence: Hydrogen and Oxygen need to be bonded in the respective order of their valencies 1 and 2.

Revised sentence: Hydrogen and Oxygen have valencies 1 and 2 respectively, and should be bonded accordingly.

9. Pay adequate attention to tense while writing sentences. In scientific and engineering text, the presentation of results should be in the past tense. However, the results of calculation and statistical analysis should be in the present tense.

Original sentence: Helium was then combined with five other gases, but it created no reaction. These results clearly show that Helium is an inert gas.

Revised sentence: Helium was then combined with five other gases. However, it did not create any reaction. The results clearly showed that Helium is an inert gas.

10. Substitute concrete and specific words for abstract words and use common and simple words instead of 'stylish' and 'buzz'

words. For example, use 'method' instead of 'paradigm' and 'best route' instead of 'optimal route'.

Original sentence: The scientist came up with a new paradigm for arriving at an optimal route.

Revised sentence: The scientist came up with a new method to get to the best route.

To summarize, the words in a sentence should be arranged to clearly convey the writer's ideas. Clumsiness has no place in scientific or engineering writing. The text must be crisp. Long sentences undermine the clarity of a passage and make easy understanding somewhat difficult.

An example of a technical passage that can be revised using the ten principles is provided below.

Original text:

A mobile phone can change your voice sounds into electrical signals, then into weak radio waves which are picked up by a transmitter receiver (TR), a part of the cellular network which changes the radio signals back into electrical ones and sends them to the local exchange. There the electrical signals were converted to flashes of laser light and sent along fibre-optic cables to the ground station, where the laser-light signals were changed to electrical ones and then to radio waves which were beamed up to Comsat in space which strengthens the signals it receives and beams them back down to another ground station two continents away where the signals were changed from radio waves into microwaves. These were beamed cross-country on a microwave link and at the main exchange, the microwaves became laser light signals again. After this conversion from light to electricity at the local exchange, the signals reach the handset and become sounds again.

The passage has four long sentences. It also has redundant words (such as 'voice sounds'). The tense also changes constantly (are, were) and there are hesitant phrases (can change).

Here is the revised version of the same passage, using the principles stated.

Revised text:

A mobile phone changes your voice into electrical signals, then into weak radio waves. The waves are picked up by a transmitter-receiver (TR), part of the cellular network. The TR changes the radio signals back into electrical ones and sends them to the local exchange. The local exchange converts electrical signals to flashes of laser light and sends these along fibre-optic cables. The ground station changes the laser light signals to electrical ones and then to radio waves, which are beamed up to a Comsat in space. The satellite receives the signals, strengthens them, and beams them back to another ground station two continents away. The signals are changed from radio waves into microwaves and beamed cross-country on a microwave link. At the main exchange the microwaves become laser light signals again, for the fibre-optic network. After conversion from light to electricity at the local exchange, the signals reach the handset and become sounds again. Your voice comes out of the handset.

The revised passage is simple and easy to understand.

It should be borne in mind, however, that sentences should neither be too long nor too short. Very short sentences will make the text seem abrupt and loosely put together. A sentence is the backbone of any communication and must also be free of grammatical errors. It should be direct, consistent, methodical, and interesting. The use of the above principles will lead to brevity in the text.

2.2 Techniques of Paragraph Writing

After words and sentences, the paragraph is the next unit of expression. There is no place for ideas of different types in the same paragraph. A paragraph itself is a part of a larger unit—the complete text—and, hence, has to fit neatly into that unit. Thus, it must have links with the preceding and subsequent paragraphs.

It might be worthwhile to make a rough chart of the paragraphs, highlighting how methodically and systematically one paragraph follows the other. It is also a good idea to make a structural design of the scientific or engineering text before beginning to write the paragraphs.

A topic sentence introduces the main idea of the paragraph. It is generally the first sentence of each paragraph and clearly provides an idea of the subject matter that will follow. It serves as a sort of introduction to the paragraph. It provides an idea of the assertions and arguments that will be communicated.

Paragraph divisions gradually unfold the assertions and arguments of the writer in stages. Each paragraph, in addition to the topic sentence, focuses on these assertions. It is then followed by sufficient proof to support the assertions made. These assertive paragraphs are the backbone of any text. To strengthen the assertions and to prove or disprove the relevant points, any number of paragraphs may be written. Of course, the law of brevity should be followed here too. Only what is essential should find a place in such assertive paragraphs. Redundancy has no place in scientific or engineering writing.

Moreover, assertions should not be overstated. Overstatements may signal that the writer is not sure of his/her assertions. A clear and complete statement of purpose of the correspondence or report, explained logically, will go a long way in establishing the writer's point of view.

Concluding paragraphs come at the end. Here the writer rounds up the main theme of the text. But there should be no re-statement of what has been said before. The concluding paragraph just states the main points of the assertions made and then forcefully points out the overall implications.

Some rules that should be kept in mind while writing are listed below:

- a paragraph should be a complete unit
- it should have a single idea

- it must contain a topic sentence which will provide a clear idea about the subject
- a paragraph should be framed so that it acts as a means of transition from one idea to another in subsequent paragraphs

Here are a couple of paragraphs illustrating the use of these techniques:

The present socio-economic development efforts have put a heavy burden on the resource base of our country. The natural resources have been indiscriminately utilized in this process to achieve short-term objectives without giving due attention to long-term environmental problems. For example, large dams were constructed for power, irrigation, and flood control, which have led to the destruction of wildlife habitat, extinction of biological species, water logging of land, sanitary and other ecological problems. Similarly, the large-scale use of chemical fertilizers and pesticides has polluted inland and coastal waters. Thousands of kilometers of roads were constructed in steep, hilly terrains which have caused major landslides and soil erosion. Industrial effluents have polluted both water and air.

Under these circumstances, the environmental conditions in the South Asian countries are likely to deteriorate further. Thus, the path of economic development that the South Asian countries have pursued so far, no longer appears tenable today.

The first paragraph has a single idea and is a complete unit. The theme of the paragraph is:

The resource base of the country has been seriously affected by the ongoing socio-economic development efforts.

This is also the topic sentence of the paragraph:

The present socio-economic development efforts have put a heavy burden on the resource base of our country.

The content of the next paragraph clearly indicates that there has been a smooth transition from the first paragraph to the second. The second paragraph, thus, begins with:

> Under these circumstances, the environmental conditions in the South Asian countries are likely to deteriorate further.

The paragraph also makes a forecast that, in the future, the conditions are likely to deteriorate further:

> Thus, the path of economic development that the South Asian countries have pursued so far, no longer appears tenable today.

It can be inferred that the subsequent paragraphs will further elaborate on this forecast.

2.3 Measuring the Clarity of a Text through Fog Index or Clarity Index

It is possible for a scientist or engineer to determine the clarity of her/his writing by using a *fog index* or *clarity index*. The index functions as an indicator of the clarity of any text. Using a formula, it determines how difficult or *foggy* the passage is. However, the fog index is *not* an indicator of how good the writing is. It is a method used to judge the crispness of any written material. The method is mechanical and objective, and works step by step. While using these steps, one must choose a sample that contains at least one hundred words to be statistically relevant. The steps are provided below.

Calculation of Fog Index or Clarity Index

The steps to be taken when the fog index needs to be measured are as follows:

Step 1. Count the total number of words in the passage.
Step 2. Count the number of sentences in the passage.

Step 3. Count the average number of words in each sentence (by dividing the number of words by the number of sentences). For example, a 100-word passage comprising 10 sentences would have an average sentence length of 100/10 or 10 words.

Step 4. Count the number of words with three syllables or more in the passage. While calculating, do not add:
- Words that are capitalized (such as proper nouns, e.g., Bangalore, Minneapolis, Jawaharlal, etc.)
- Words that are a combination of short easy words, such as, undertake, bookkeeping, anyone, multinational, etc.
- Verb forms whose number of syllables has increased by the addition of -ed, or -es or -ing, such as credited, etc.

Step 5. Add the last two numbers (derived in Steps 3 and 4 above).

Step 6. Take four-tenths of this total.

The final result determines the fog index or clarity index. Here, low scores are desirable and imply high clarity of writing.

The fog index, originally developed by Robert Gunning, along with subsequent modifications, is considered the most reliable method for determining how difficult or easy a passage is to read and understand. The ideal fog index of a passage is between 8 and 9. A passage with a fog index higher than 15 must be rewritten to lower the index and improve clarity.

An example of a passage with a very high fog index is:

Short, unambiguous words are the bricks—the foundation— that build up the structure—the sentences, paragraphs, and ultimately the whole paper—and so maximum attention and importance is to be imparted to the selection of words. It has to be remembered that scientific writing has a definite objective in view, its most important purpose being to communicate something by means of a signal—the vocabulary that is being used. This signal should be intelligible, transparent, and unambiguous. Punctilious adherence to these instructions will make a manuscript more comprehensible. Deficiency in clarity inhibits the refinement of the technical content of the paper.

The fog index of this passage will be determined as follows:

1. This is a passage of about 100 words. The passage has five sentences. Hence the average number of words in a sentence is 100/5 = 20.
2. There are 30 words with more than three syllables in this passage.
3. On adding steps 1 and 2, we get 50.
4. Four-tenths of this total is 20.

Therefore, the fog index of this passage is 20, which is too high.

It is possible to convey the ideas expressed in this passage by using the ten principles stated earlier. This will make the passage much simpler, and will automatically bring down the fog index. It is important to ensure, however, that when simplified, no important point is omitted.

The simplified version of the passage is provided below.

Short simple words are the bricks—the base. The sentences—the paragraphs and in fact, the whole structure—depend on these words. Simple words are not only clear, but are also more forceful than big words. Hence, utmost care has to be taken while choosing words. A scientist writes with a certain purpose in mind—to convey a message to fellow scientists by means of a signal. The words he/she uses are this signal. This signal must be clear. Lack of clarity spoils the technical content of the paper.

To find out the fog index of the revised passage, we follow the same sequence of steps:

1. The number of words in the paragraph is 80. The number of sentences in the paragraph is eight. Hence, the average number of words per sentence is 10.

2. The number of words with three syllables is 5.
3. On adding the last two numbers we get 15.
4. Four-tenths of this total is 6.

Hence the fog index of the passage is only 6, much lower than the previous 20 for the same passage written in a more complex manner.

The easiest way to keep the fog index down is to write short sentences. The average length of the sentences should be limited to around 15 words per sentence. But it is not necessary to make every sentence short and of the same length. In fact, this might lead to monotony. The length of sentences should be varied according to the requirements. The fog index could even be raised, if required, by joining sentences. This could not only smooth out the choppiness or abruptness, but would also help reflect a better relationship between ideas. This is an essential feature of communication for engineers and scientists.

While the fog index or clarity index provides a measure of the crispness of the communication and clarity of the language that is used in the text, it denotes nothing more. Knowledge of the technique of sentence and paragraph writing, keeping the fog index in mind, is essential for those who wish to write clearly.

3

Writing Scientific and Engineering
Papers

In an age when
publish or perish is the 'mantra',
writing scientific papers
is the creed of the day.

THE intent of a scientific or engineering paper is to publish findings to draw the attention of the scientific or engineering community. Scientists and engineers are aware of the visibility that their findings will receive if they are published in a reputed scientific or engineering journal. It is, therefore, quite natural for them to want to see their new research findings published. For this, they have to write a research paper incorporating a detailed statement of their investigations and researches. This task, however, is not easy. As authors, they have to adhere to the basic principles of paper writing. Scientists and engineers must be aware of the technicalities involved in writing a scientific or engineering paper; and should also have a comprehensive knowledge about the *format* of a scientific or engineering paper.

Format refers to the general plan of organization of the material and findings that a scientist or an engineer may want to include in her/his paper. A scientist or an engineer, in the course of research, arrives at many new findings. Once these are ready, a structure is needed, on which to weave a story for the broader community.

The author needs to begin the story with the goal of the investigation, which may include specific problems addressed or certain hypotheses that were tested. Mention has to be made of the methods adopted and the material used in the study. With the help of tables and figures, the key results of the investigation have to be presented. Finally, she/he has to explain the significance of these findings and plans for future work.

In a scientific or engineering paper all these cannot be presented in an ad hoc manner. What is needed is an accepted pattern. The *format* is the technical pattern that makes the structured weaving of the story possible. Every scientific or engineering paper has to adhere strictly to the basic technical structure and format of the paper. It is imperative that all relevant information regarding the subject of study is incorporated in the appropriate place in the paper. A badly structured paper has an inherent inertia, which impedes the step-by-step progress of the paper.

The targeted format of a scientific or engineering paper should be as follows:

- title
- abstract
- introduction
- materials and methods (experimental details or theoretical basis)
- results
- discussion
- conclusion
- references (or literature cited)
- acknowledgements
- appendix

This general format has become standard because it is suitable for most reports of original research. It is logical and easy to use. The reason it accommodates most reports of original research is that it parallels the scientific method of deductive reasoning. It defines the problem, creates a hypothesis, devises an experiment to test the hypothesis, conducts the experiment, and draws conclusions. Furthermore, this format enables the reader to understand quickly what is being presented and to find specific information easily. This ability is crucial today, more than ever before, because all professionals must read much more material in a limited time. Hence, most scientific or engineering publications require revisions of manuscripts that are not in the requisite format.

The format of a scientific or engineering paper has grown out of years of developing traditions, editorial practice, and scientific ethics. It was prescribed by the American National Standards Institute (ANSI) years ago. Whatever the technical discipline, all research papers follow this general format, with minor deviations. But a target journal might have its own criteria and therefore, it is always prudent to refer to the stipulations provided in the 'Instructions to Authors'.

To summarize, adhering to a format has certain distinctive advantages:

(a) The logical structure of the format helps gradual and step-by-step unfolding and display of the contents of the paper.

(b) It helps an author to effectively communicate the scientific findings to the concerned community in a logical, uniform, and methodical manner.

(c) It helps the reader to understand the author's viewpoint from different angles.

(d) The format enables the reader to identify and concentrate on a section.

(e) It helps the reader to get a quick view of the sections of interest.

The following table describes the key aspects of each section of the format. Each of these is described in more detail in subsequent sections.

**Table 3.1: An Overview of the Format of a Scientific/
Engineering Paper**

Title	It should be succinct, have keywords, and convey the message.
Abstract	The abstract is the essence of the other sections. Its main objective is to inform the readers about the goal of the experiment, the result, and the significance of the findings. The abstract must be self-contained.
Introduction	The introduction defines the subject of the paper, explains the conflicts and gaps in existing knowledge and the purpose of the research undertaken. It should mention what problems were investigated and what hypotheses were tested. It should clearly establish the significance of the present work.
Materials and Methods	This section should describe what was done and how. It should provide an idea about the materials and methods used.
Results	The results section should present the key results in a logical sequence with the help of tables and figures. Figures and tables should be self-explanatory.
Discussion	This section should be devoted to the interpretation of the results vis-à-vis the original objectives or hypotheses. Here the theoretical as well as the practical implications of the work should be thoroughly discussed. Possible directions for future work on a similar subject should also be indicated here.
Conclusion	Conclusions must conclude. The main function of this section is to summarize the intrinsic value of the research.
References or Literature Cited	The source of any information borrowed from a paper or book must be indicated here.

Acknowledgements	Financial support from a grant-giving agency or others who provided technical or material support must be acknowledged here.
Appendices	Supplementary information—useful, but not essential, or of interest to only a subset of readers—is included here.

All sections of the format, along with their contents, have been discussed next.

3.1 Title

The title of the paper should generally have less than ten words. Its main purpose is to reflect the factual intent of the paper. It must be specific and informative. It should express succinctly what the paper is about. Its objective is to convey to the readers the broad idea and results. A good title achieves this in the least possible time and space.

An appropriate title has keywords that would be of help to researchers. Judicious use of keywords helps interested scientists working in the same field to locate the paper. Indexing and abstracting services depend on the accuracy of the title – keywords extracted from which might be useful for references and computer searches. In a good title, redundant expressions like 'A Study of ...' have no place. In short, a title should be such that it draws the reader's attention and interest at once.

The following list summarizes the key tenets of the title.

- A table should not have more than ten words;
- must be specific and informative;
- should clearly identify the factual intent of the paper;
- must have keywords which facilitate computer searches;
- must have no redundant words;
- should be clear, precise, and forceful; and
- need not follow formal grammatical structure.

An example of an inappropriate title is:

> The study of some novel multiplier design methods that can be effectively used to achieve very high frequency designs

A quick look at this title reveals its basic inadequacy. Some of these are:

- It is long (19 words).
- It has a number of redundant words (as stated earlier, it is not necessary to follow a formal grammatical structure in titles).
- It is not forceful (phrases like 'The study of some...' lead to a weak beginning).
- It is not precise and specific (it does not state how high the frequency is, or what novel multiplier design method has been used).

This title can be revised as follows:

> Crossing the Giga-Hertz Limit with Adaptive Multiplication

The characteristics of this title are:

- It is short (7 words) and has no redundant words.
- It is forceful (a beginning like 'Crossing the...Limit' gives it an intensity).
- It has appropriate keywords ('giga-hertz' and 'adapative multiplication' lend themselves to useful computer searches).
- It is very precise and specific (it clearly states that the frequency is giga-hertz and the novel multiplier design method is adaptive multiplication).

Thus, it adheres to all the principles of a good title.

Another example of an inappropriate and appropriate titles is:

Original Title: A new method to enhance the clarity of photographs derived from moderate accuracy camera phones

Revised title: Differential Imaging for 8-Megapixel Camera Phones

3.2 Abstract

Immediately after the title comes the abstract. The abstract is a succinct summary of the paper. Its objective is to help readers to identify the basic content of the paper quickly and accurately. Hence, it should focus on the main intent of the paper.

It is a good strategy for the author of a scientific or engineering paper to write the abstract after writing the rest of the paper. This is because the abstract is the essence of the other sections. The author should collect important material from different sections of the paper and incorporate them concisely in the abstract.

In any scientific or engineering text, the 'introduction' section provides a summary of the problem that has been investigated. It also indicates the gaps in current literature that have been closed through the research. Following this, the 'materials and methods' section provides an idea about the basic methodology adopted in the scientific endeavour. Next, the 'results' section provides information about the key quantitative observations made in the study. Lastly, the 'discussion' section states the implications of the new findings from the research and the experiments conducted.

However, while absorbing key points from each of these sections for the abstract, no particular point should be over-emphasised. It is also critical to note that any information that has not been discussed in the paper cannot be communicated in the abstract. There is also no place in the abstract for any figure, table, or any sort of illustration. Nor should there be any reference to other literature or lengthy background information.

The abstract should be concise, crisp, compact, and written in one paragraph of about 200 to 250 words. It must be self-contained as it is often distributed along with the title separately from the paper. The abstract should focus on the main objective of the paper. The main objective is to let the readers know what the goal of the experiment was, what the results were, and what the significance of the findings is. The abstract plays a key role in the organization of the entire paper. The abstract is followed by the introduction, which ushers in the main body of the paper.

An example of a well-written abstract is from a technical paper titled 'Allocation and Binding in Data Path Synthesis Using a Genetic Algorithm Approach', authored by C.A. Mandal, et al.

Abstract: A technique for allocation and binding for data path synthesis (DPS) using a Genetic Algorithm (GA) approach has been developed. The proposed genetic algorithm uses a non-conventional crossover mechanism, relying on a novel force directed data path binding completion algorithm. The proposed technique has a number of features such as acceptance of some design parameters from the user, use of a bus-based topology, use of multi-port memories and provision for multi-cycling and pipelining, among other features. The results obtained on the standard examples are promising.

Keywords: Data Path Synthesis, Binding, Multi-port Memory

In contrast, an example of an improperly written abstract is:

Original abstract:

Semiconductor solutions in the automotive world of today are extremely complex and are the brain behind controlling all sophisticated operations in the environment in which it is working. This requires heightened focus on reliability before the product is released to the market. Most of the issues in automotive products stem from lack of verification of extreme and outlier conditions. This paper introduces the theme automotive-grade quality and reliability and provides techniques for the system, hardware, and software developer to successfully architect, develop, and execute it to verify a new device that guarantees near-zero failure modes in a million. Although my experience is in automotive electronics, the techniques here are fit for any other domain that is very sensitive to high quality and reliability yardsticks in mission-control applications.

This abstract can be improved by:

- removing redundant expressions, such as, 'This paper introduces...', '...environment in which it is working', 'extreme and outlier', etc.
- avoiding awkward expressions, for example, phrases such as 'Although my experience is in...' should be avoided.
- using simpler substitutes for wordy phrases, for example, the word 'generic' can be used instead of the phrase 'fit for any'.
- adding keywords, such as, automotive-grade quality, reliability, etc.

Using the above techniques, the abstract may be revised as follows:

Revised abstract:

Semiconductor solutions for automotive-grade applications push the envelope in terms of quality and reliability. Most issues encountered today arise from an inability to simulate extreme conditions. This paper has proposed a generic approach that integrates the systems, software, and hardware aspects to reliability. The results obtained are significant and promise to lead to near-zero defects per million.
Keywords: Automotive-grade quality, Reliability

The seven features of an abstract have been listed below.

1. An abstract must be self-contained.
2. It should be printed as a single paragraph.
3. While writing the abstract, the main ideas should be collected from each section following a cohesive sequence.
4. Information that has not been mentioned and discussed in the paper should not be included in the abstract.
5. As the abstract refers to an investigation that has been completed, it should be written in the past tense.
6. There should be no tables, figures, or illustrations
7. The abstract should contain keywords. The indexing and abstracting services depend on the keyword system.

3.3 Introduction

The most important function of the 'introduction' is to establish the main subject under investigation and bring out the significance of the current work. It provides sufficient background information to enable readers to develop an interest in the subject. It gives the readers an idea of the reasons that prompted the author to undertake this particular study. It showcases the problems that were investigated and the hypotheses that were tested. In short, the introduction defines the subject of the paper and clearly explains the technical purpose of the research.

A good introduction is expected to provide the answers to the following questions:

- What knowledge already exists in the domain of the paper?
- Are there any conflicts or limitations in that knowledge?
- What gaps prompted the study and what is its importance?
- Why was this kind of experiment or experimental design chosen?
- What problems were investigated?
- What hypotheses were tested?
- Did the results tally with the hypothesis formulated?
- Have the gaps/conflicts been successfully demonstrated?
- What are the suggested solutions?
- How are the solutions different from/superior to existing ones?
- How has the paper been structured (by section)?

The introduction, therefore, should begin with the full background and contain all relevant general information. This should be followed by a description of the specific problems and the methods adopted to solve these. If a new methodology has been used, it should be detailed clearly. Finally, the introduction should end with information on how the paper has been structured by section.

We will illustrate this gradual narrowing of scope, from generic background information to the specific problem statement, and

ending with the section-structure, by referring to an excellent excerpt from a scientific paper titled 'Two-photon dissociation of
HD^+ in two-frequency laser fields in the presence of two intermediate resonances'.

The authors, Banani Dutta and S.S. Bhattacharyya, begin by
informing the reader about the current understanding of scientists
in the general area of multiphoton processes. Through several
references, the maturity and results from prior experiments in this
field are first stated in detail:

> Over the last decade or so our understanding of multiphoton
> processes has advanced remarkably. The progress, achieved
> to a large extent by many theoretical studies, has been most
> significant for the case of multiphoton ionization of atoms.
> A large number of theoretical works have also explored novel
> features to be expected in the behaviour of systems in intense
> electromagnetic fields by considering models with various
> interesting configurations of resonant levels.
>
> Examples of such works can be obtained from a recent
> issue of *Journal of Optical Society of America* (1990) and also
> the review article of Knight et al. (1990). Empirical varia-
> tions of the model parameters enabled the authors of these
> works to investigate a wide range of possible dynamical
> behaviours of realistic atomic systems under the influence of
> one or more intense electromagnetic fields.
>
> On the other hand, theoretical studies on multiphoton dis-
> sociation of small, simple diatomic molecules in intense laser
> fields have begun comparatively recently and have not yet
> reached the same stage of maturity. Nevertheless, in recent
> years such studies have also started to yield interesting
> insights into the multiphoton absorption of small molecules.
> In particular, we would like to mention the application of
> resolvent operator techniques (Goldberger and Watson, 1964;
> Faisal 1987), originally applied to the problem of two-
> photon ionization of atoms (Lambropoulos, 1974; Beers and

Armstrong; 1975, Georges and Lambropoulos, 1980; Raczynski and Zaremba, 1987) and strong field autoionization (Lambropoulos and Zoller, 1981; Kim and Lambropoulos, 1984; Bachau, et al. 1986), to problems of multiphoton ionization of molecules (Domcke, et al. 1988; Rai Dastidar and Lambropoulos, 1984) and photodissociation (Lami and Rahman, 1981, 1982).

We had also previously used this technique for studying resonant two-photon dissociation (TPD) of HD^+ in a single frequency and two frequency field.

They follow this broad showcasing of the work done in the domain by stating the gaps in the existing research. This sets the stage for the authors to highlight the reasons behind the proposed research. The following excerpts from the paper illustrate the above:

…However, the (prior) study was restricted to the case of a single resonance with the initial state.

In this work, we extend the study to the case when two different bound levels are resonantly coupled to the initial level by two fields of different frequencies….

Thus in this work we have calculated TPD rates and probability as well as the photofragment energy spectrum of HD^+ under the action of two laser fields.

Finally, the introduction ends with information on how the sections have been structured. Any deviation from the standard structural norm of presentation is also highlighted here. The following excerpts from this paper illustrate this.

In section 2 we give the general formulation of the problem and in section 3 we briefly discuss the way the calculations have been performed, mentioning in the process, the main assumptions and approximations used. Finally, we report sample results for some selected combinations of intensities and frequencies.

It is evident from this excerpt that it is important to ensure that all aspects of the introduction are presented logically, methodically, and unambiguously.

To sum up, the introduction should

- clearly define the state-of-the-art in the chosen domain including any limitations in the current research.
- clearly state the problem being investigated and how the paper closes the gaps stated.
- clearly elaborate the merits of any new method if any novel technique has been used.
- clearly state how the rest of the paper has been structured.

3.4 Materials and Methods

The materials and methods (or procedure) section should clearly describe what was actually done. This section should be written in a narrative, paragraph format—not just as a list of numbered steps. At this stage it should not include any results. It is necessary to use subheads when more than one experiment has been made. Besides the subheads, each experiment should be indicated as a unit. It is always better to provide necessary quantitative details, like how much, how long, etc., to describe the experiments. This section should clearly explain how the data was analysed. The statistical procedures followed for analysis also need to be explained. This includes the probability level at which the result was determined. It must also provide information on errors of measurements for all measurements provided in this section. The experimental design should be clearly described. It is necessary to include the hypotheses tested. Controls, treatments, and variables measured should also be mentioned. The number of times the experiment was repeated must be mentioned along with the findings for each.

A drawing of the experimental apparatus adds to the richness of this section. Any modifications of equipment or the equipment constructed specially for the experiments should be carefully described. The materials and methods section should specify all experiments leading to the results. It should be done so that there is an uninterrupted logical flow of the description being given. It should provide answers to the following:

(a) How the study was conducted?
(b) What equipment was used?
(c) What procedures were followed?
(d) Where and when the work was done (especially in field studies)?
(e) What was unique about the method?

Here is an example from a paper titled 'Molecular pendular states in intense laser fields' by G. Ravindra Kumar, et al. to indicate how this section is to be written while adhering to all the principles.

Experimental Method

The angular distributions were measured using a crossed-beam apparatus that has been described in recent reports on the competition between field-induced molecular dissociation and ionization and only the salient features are described here. Focused, linearly polarized 35-ps. 532-nm pulses (10-Hz repetition rate) from an Nd:YAG laser (where Nd:YAG denotes neodymium-doped yttrium aluminium garnet) interact with an effusive molecular beam in a field-free region and the resulting ions are mass analyzed by a quadrupole filter.

Ion angular distributions were measured by rotating the laser polarization vector with respect to the quadrupole filter's axis: the final polarization state was selected using a linear polarizer, which was rotated in steps of 2°–4°. The linearly polarized laser radiation was passed through a half-wave plate to rotate the polarization vector and to ensure that the final energy of the laser pulse was kept constant. The shot-to-shot reproducibility of the laser was ±5%.

The unique feature of the present experiment is the complete absence of an electrostatic field to extract the ions into the mass filter; almost all earlier measurements have all utilized time-of-flight ion analysis techniques in which large extraction fields (of the order of 600 V cm^{-1}) are mandatory. Such fields result in distortions of the measured angular distribution functions, particularly those pertaining to multiply charged ions. In our case, only those singly charged ions whose initial velocity vector lies within the acceptance angle set by the entrance aperture of the mass filter are detected; the best angular resolution was achieved ±1.8°. Adequate precautions are taken to avoid saturation effects by keeping molecular beam densities low and restricting laser beam energies to values that avoid saturation of the detector.

An analysis of the above section brings out the following attributes of the method:

- The equipment used ('crossed-beam apparatus') has been clearly stated and referenced ('...described in recent reports on the competition between field-induced molecular dissociation and ionization ...')
- The way the study was conducted has been succinctly described ['Focused, linearly polarized 35-ps. 532-nm pulses (10-Hz repetition rate) from an Nd:YAG laser interact with an effusive molecular beam in a field-free region and the resulting ions are...analyzed by a quadrupole filter'].
- The procedure followed for the measurement has been stated in a manner that can be replicated easily ('Ion angular distributions were measured by rotating the laser...in steps of 2°–4°. The linearly polarized laser radiation was passed through a half-wave plate to...ensure that the final energy of the laser pulse was kept constant. The shot-to-shot reproducibility of the laser was ±5%... Precautions are taken to avoid saturation effects by keeping molecular beam densities low and restricting laser beam energies to...avoid saturation of the detector').

- The uniqueness of the experiment has been described well ['The unique feature of the present experiment is the complete absence of an electrostatic field to extract the ions into the mass filter; almost all earlier measurements have all utilized time-of-flight ion analysis techniques in which large extraction fields (of the order of 600 V cm^{-1}) are mandatory'].

In some cases, it is better to describe the method adopted using flow-charts and tables. The section given below, from a paper presented at the 2002 VLSI Design and Test Conference by C.P. Ravikumar and Rahul Kumar, illustrates this.

Experimental Flow

...In our own work (see Fig. 3.1), we used Design Compiler for synthesis and TSMC 0.18 μm technology... Leakage power and the number of cells used by the design are extracted...through Perl scripts... The experiments were performed for three libraries: slow, fast, typical... (see Table 3.2).

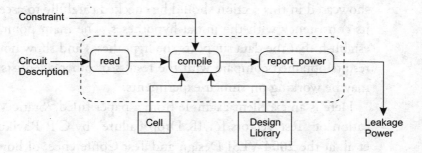

Fig. 3.1: Leakage Power Estimation

Table 3.2: Operating Conditions for 0.18 μm Technology

Library Name	Voltage (V)	Temperature (°C)
Slow	1.62	125
Typical	1.8	25
Fast	1.98	0

3.5 Results

The most important section—the core—of the paper is the 'results' section. The results comprise the new knowledge—the new findings that are being presented to the scientific or engineering community. It is an overview of what has been accomplished by the author(s). The main function of the results section is to clearly present the key results in a simple, transparent, and logical sequence with the help of tables and figures.

What is the key result? At the very beginning of the study, the scientist or engineer poses certain testable hypotheses. Subsequently, these hypotheses are tested through various types of observations and experiments to determine their validity. The answer obtained from these tests is the key result.

The results section should be written with sufficient care; there should be no interpretation at this stage. There may be brief summaries of the statistical analysis. It is important to ensure that instead a series of raw data, only meaningful and representative data should be used – with major focus on *analysis* of the data. The data showcased in this section should be checked carefully to ascertain its consistency with the initial hypotheses. The main point is to establish that the data supports the hypotheses and show how the results obtained compare with the results of other scientists who may be working on similar experiments.

Here is an excellent example from a paper titled, 'Static Verification of Test Vectors for IR Drop Failure', by C.P. Ravikumar, et al. at the 2004 VLSI Design and Test Conference, of how this section can be written effectively.

Results

We tested the flow presented in the previous sections on two industrial designs, called Chip-A and Chip-B. Chip-A is a memory-intensive design with 70K gates and 64KB memory, with a single scan chain. Chip-B has around 3 million gates and about 840KB of memory, and has 7 scan chains.

Figure 3.2 shows the results of the vector short-listing technique proposed here. We plot the toggle counts for the initial 19 test vectors.

The test vectors passed the latter flow, but the TestRail flow caught the failing vector, as shown in the AstroRail plot of IR-drops (Figure 3.3(a)). We see that central portion in the chip is colored red, meaning the IR drop in this area exceeds the margin. To contrast, we also created functional test vectors which exercised only subsets of the memories... The TestRail flow did not find any problem in such functional vectors (see Figure 3.3(b)).

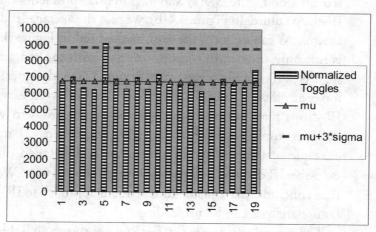

Fig. 3.2: Toggle Counts for Chip-B for 19 Vectors

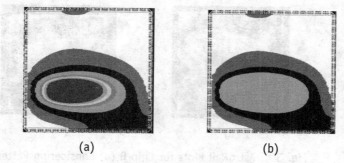

(a)   (b)

Fig. 3.3: Plots of IR-Drops (a) for Failing Vector
(b) for Functional Vector

Figure 3.4 shows the results on the vector short-listing technique proposed here. We plot the toggle counts for the initial 19 test vectors. The toggle count for Vector 5 is unusually large, larger than the average $+3*$ standard deviation, making it a...candidate for shortlisting. Vector 5...failed on this chip, and is mentioned in Figure 3.4(b). Clearly, the shortlisting technique saves significant amount of run-time by avoiding further analysis.

Figure 3.4 shows the AstroRail plots of the failing test vector for Chip-B. In Figure 3.4(a), the IR-drops are averages, calculated in a pattern-independent fashion. We see that the red area of Figure 3.4(a) shows potential problems due to IR-drop failure. In Figure 3.4(b), we plot the average IR drops again... As can be seen, the area marked red in Figure 3.4(b) is not only smaller, but has no overlap with the red area in Figure 3.4(a). This means two things: (1) A designer who looks at plot 6(a) and fixes the power rail to avoid excessive IR-drops may not really address the actual problem while still consuming costly silicon real-estate; and (2) in the example of Chip 6(b), the critical path that failed due to excessive IR drop indeed passes through the red area. We can thus conclude that a vector-independent approach to IR-drop failure analysis can be misleading.

The typical run-times for TestRail are shown in Table 3.3 for different phases of the total flow. As can be seen, for a

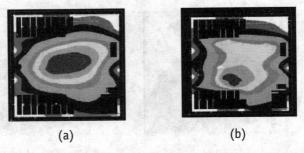

(a)   (b)

Fig. 3.4: AstroRail Plots for Chip-B (a) Considering Pattern-independent Averages (b) Considering Pattern-dependent Average IR-drops

Table 3.3: Run-times for different Phases of TestRail

	Vectors short-listed	Vector short-listing	Toggle count propagation	Power rail analysis	Critical path analysis
Chip-A	5	< 1 min	~ 5 min	~ 5 min	~ 20 min
Chip-B	80	~ 5 min	~ 5 min	~ 2 hrs	~ 1.5 hrs

large design like Chip-B, the vector shortlisting can proceed at the rate of about 1 vectors/4sec. The toggle count estimation was done for the one shortlisted vector in case of Chip B, and this step took ~5 min. This clearly brings out the benefit of the vector shortlisting proposed in this paper; had we performed toggle count estimation for all the 80 test vectors, the total run time for even the toggle count estimation phase would have exceeded 3 hrs, not to speak of the run-times during Astro-Rail and timing analysis phases of the flow.

This example of the 'results' section uses multiple figures and tables to present the experimental data and analyse these. It also tests the original hypothesis stated at the beginning of the paper and states conditions where these were found valid.

The Role of Tables and Figures in the Results Section

The preparation of tables and figures is of immense importance for the presentation of the key results. This is clearly borne out by the previous example. When all the data has been carefully analysed, the tables and figures should be prepared. All the tables and figures are to be arranged in their proper sequence so that the findings may be intelligibly and methodically expressed. Attention has to be given to the following points:

(i) Tables and figures should have assigned numbers. These numbers are to be strictly given in the sequence that they are referred to in the text.

(ii) A brief description of the results being presented should be given in the form of a legend with each table and figure. The table legends should appear above the table and the figure legends below.

Pictures are more expressive than words. Figures should be used wherever necessary to communicate specific technical results. But figures should never be placed in an ad hoc manner. They should be placed immediately after they are mentioned in the text. Otherwise, they may be placed at the end of the paper. Each figure should have a number and title, so that it may be referenced from the text. Readers should be able to understand the figures independently without the help of the text. All the tables and titles should be provided in one contextual framework.

Effective scientific and engineering communication depends to a great extent on the writer's ability to synthesize huge amounts of data. Representing data using charts and tables increases clarity significantly.

Charts are a very succinct way to communicate data. These can be of various types such as bar charts, pie charts, bubble charts, scatter diagrams, etc. In each of these charts, there can be variants for:

- showing the data values in the chart
- charts without grids
- charts with linear, logarithmic, or other scales

A description of different types of charts and tables is provided in Chapter 5.

3.6 Discussion

The 'discussion' section is devoted to the interpretation of the results vis-à-vis the original objectives or hypotheses. The main objective of this section is to explain the salient features of the new findings, which lead to the results. The author should clearly mention if he/she agrees or disagrees with any study or research area previously explored.

The fundamental questions addressed in this section are:

- What are the actual implications of the problems being investigated?
- Will the investigations and experiments and the subsequent results conclusively prove the testable hypotheses?
- In case the findings differ from the results of other researchers working in the same field, what would be the author's view? Could it be due to an error or a different perspective in the author's experiment or in a prior work?
- What are the conclusions that can be drawn from the results of the experiments conducted?

In the 'discussion' section, the theoretical implication as well as the practical implication of the work has to be thoroughly discussed. And lastly, possible directions for future research on similar subjects should be provided. The discussion section should end with a succinct summary about the significance of the study.

A good example of the discussion section from a paper titled 'Effect of Laser Polarization on X-Ray Emission from Ar_n (n = 200 – 10^4) Clusters in Intense Laser Fields', authored by Kumarappan, et al. is given below.

We have studied the x-ray emission from argon clusters in intense laser fields as a function of the ellipticity...The x-ray emission does not vary with the ellipticity of the laser at any stagnation pressure or incident laser intensity. These results suggest that the electron energy distributions...for x-ray emission do not vary with the ellipticity of the driving field.

Our findings have to be placed in the context of the nanoplasma model. The high degree of collisionality that is inherent in plasma situations does not automatically imply that the plasma dynamics is simple and predictable. For instance, multiple inner-shell vacancies observed in the experiments of Shroeder et al. remain difficult to rationalize in terms of thermalized electron energy distributions... There is mounting evidence that polarization-dependent properties in solids...play a major role in determining the overall dynamics [12] and that information on the initial polarization state of the incident radiation is not necessarily 'lost'. It appears somewhat surprising that our results can be rationalized on the basis of extrapolations of recent Monte Carlo simulations of electron energy distributions in high-density gases [14]...

3.7 Conclusion

A 'conclusion', as the term suggests, comes at the end of a paper and is the section wherein all previously discussed issues are tied up. Its main function is to provide an overall insight into the intrinsic value of the thesis presented in the paper. A conclusion must provide a clear idea of the impact of the findings arrived at and incorporated in the text. Everything stated in the paper earlier should support the conclusion and smoothly and naturally lead up to it. This would be a very appropriate end to a scientific or engineering paper.

Here is an example of a well-written conclusion from the paper 'Molecular Pendular States in Intense Laser Fields', authored by G. Ravindra Kumar, et al. from the Tata Institute of Fundamental Research.

We have demonstrated the formation of aligned states of thermal-energy triomatic molecules in intense laser fields. Pendular motion manifests itself in the measured shapes of the angular distributions of dissociation products formed as a result of molecular field interactions. The results of our dissociation dynamic model, using a LEPS potential energy surface, qualitatively reproduce the features of our data, the measured heart-shaped angular distributions are reproduced.

The alignment that is measured fits in with the theory of Friedrich and Herschbach and their interpretation of such alignment due to pendular states. We believe our measurements to be most direct evidence for pendular motion of molecules in intense laser fields.

The section above brings out the overall theme of the research and its impact.

In contrast, the following conclusion is a poorly written one.

It is very clear that each of the methods suggested in this paper for packaging efficiency are very critically valuable and hence all the measures suggested, either minor or major, when adopted, would lead up to contribute considerably towards the ultimate goal—reducing the cost.

This conclusion does not provide any insight into the findings and the specific impact of the research. It is also long-winded and lacks crispness.

3.8 References

It is imperative to indicate the source of information referred to in a paper or report. Thus, the source of any information that does not result from the experiments conducted by the author must be cited in a paper. The references (or literature cited) section is not the same as a bibliography. A bibliography is just a list of references, which may not have been cited in the paper.

There are two systems of referencing: The APA style guide and the Harvard referencing style. The former, the APA style, is based on the Publication Manual of the American Psychological Association, 5th Edition, 2001. This system is primarily used in the Faculty of Health and similar organizations.

The most widely used system of referencing in scientific and technical journals is the Harvard referencing style. It is also known as the author/date system. Here, the citation in the body of the author's text should contain two essential points:

    (a) the family name of the author(s)

    (b) the date and year of publication

An example of in-text citation is provided below:

> Effective communication is the basic requirement for leadership.
> (Bennett and Corrigan, 1981)

It is also possible to make the name of the author(s) a part of the sentence. This is illustrated below:

> Bennett and Corrigan (1981) state that effective communication is the basic requirement for leadership.

However, the in-text citation provides the information briefly. There should also be a way to identify the rest of the information. The list of references does this function. The reference list comes at the end of the paper and provides all the information about the material cited in the text.

References should be listed in alphabetical order according to the author's surname or last name. If there are two authors, both names should be cited. In case there are more than two authors, only the first author's name should be cited, followed by the expression 'et al.' (Latin—et alia; meaning 'and others'). Sources not actually cited in the text of the paper need not be included in the literature cited section. If one refers to a number of articles by the same author, these should be arranged chronologically according to the date of the publication.

In short, the literature cited section should contain the name(s) of the author(s), name of the journal or book, title of the article, year of publication, and name of the company that published the book. 'Double titling' or repeating of one title should be avoided. For example, in place of Dr Swagata Ghosh, Ph.D, one should write, Swagata Ghosh, Ph.D, because the degree Ph.D represents the title 'Dr'. Likewise, just Dr Swagata Ghosh could be written.

3.9　Acknowledgements

An 'acknowledgement' gives credit to or thanks those who helped in the research. Help may have been provided through advice, suggestions, comments, or monetary support. Giving credit for financial support to the grant-giving agency is essential.

Here is an example of how an acknowledgement should be written. This is excerpted from a paper published in *Physical Review* by V. Kumarappan, et al. (2001):

> We thank G. Ravindra Kumar and S. Banerjee for their help and many useful discussions. We also benefited from discussions with D. D. Bhawalkar and the late T. K. Dastidar. Partial financial support from the Department of Science and Technology for our high intensity femtosecond laser system is also acknowledged.

3.10　Appendices

Appendices contain information which could be presented in the main body of the paper in detail, but which might be of interest to only a subset of scientists working in the same field. However, only appendices referred to in the main body of the paper should be included. Appendices should reference all mentioned material, which is not the original work of the author. Appendices include supplementary information that is useful, but not essential.

4

Technical Guidelines for Communication

*Technical guidelines are
the pathfinders
that pave the path for
proper presentation.*

THE earlier chapters elaborated on the need for communication to be concise, precise, and error-free. Apart from these, there are some unique technicalities that are part of scientific and engineering communication.

These technicalities may be part of

- any general technical communication amongst the scientific or engineering community; and
- communication of key scientific or engineering results to a broader technical community.

Each of these is described in the sections below.

4.1 Generic Technical Exchange among the Scientific or Engineering Community

While communicating with their peers, engineers or scientists must take note of the following.

Hyphenated Words

If the first word is used as an adjective, no hyphen is necessary—as in the phrase 'first draft'. If the first word is a noun in compound adjectival phrases, then it must be hyphenated. Some examples of such phrases are: 'rock-bottom price of the product', 'short-term view of the economy', 'well-known author', 'middle-class values', etc. If the second word is a gerund (or verbal noun), that is, the present continuous tense of a verb, then there is no need to hyphenate. An example is the phrase 'earth shattering invention'.

Endings to Verbal Form (-ment, -dge, -ise, -ize)

When a verb ends with –dge, retaining the final 'e' is preferable when both spellings are in use. For example, acknowledge: acknowledgement, judge: judgement, etc. However, 'acknowledgment' and 'judgment' are also commonly used in technical and legal communication today.

Similarly, when both –ise and –ize are possible verbal endings, –ize is preferred. For example, crystallize, characterize, hyphotize, etc.

Use of the Apostrophe

An apostrophe should be used only to indicate possessives. For example: Moore's Law. However, in plural forms, the apostrophe must not be used. For example: 1990s, VIPs, etc.

Fractional Numbers

Fractional numbers are considered to be plurals. Thus, they should be written as 'one metre', but '1.5 metres'.

Integers

Integers less than ten are spelled out. For example, five miles. Integers larger than ten and fractional numbers are written in Arabic digits, that is, as 15, 8.5, etc.

Articles before Symbols

In scientific or engineering communication, deciding which article to use before a symbol can prove confusing. The rule here is the same as in general writing. The use of an article before a symbol depends on the pronunciation of that symbol. In the case of a symbol which begins with a vowel sound, 'an' must be used; in other instances, a symbol should be preceded by 'a'.

The Oxford Comma

A comma, referred to as the Oxford comma, should be used before 'and'/'or' when three or more items are referred to. For example: 'copper, iron, and zinc'. Similarly, 'synchronous, asynchronous, or both'.

Abbreviations

When abbreviations are used, the full form of the abbreviated phrase must be written in the first instance, followed by the abbreviated form within brackets. Subsequently, the abbreviated form can be used in the text. For example, in a text the first reference should be Indian Institute of Technology (IIT), while subsequently just IIT could be used.

Full stops should not be used in abbreviations that consist of initial capitals. For example, IIT, MIT, USA, GDP, etc. However, full stops should be used when the abbreviation consists of lower-case letters and when the abbreviation does not end with the last letter of the word. For example: a.m., p.m., Prof.

Expanding Forms

Forms like 'don't' should not be used; it should be used in its full expansion, that is, 'do not'. Similarly, 'and' should always be used in text, instead of ampersand (&).

Scientific or Engineering Terms

All technical names must be italicized. Likewise, all foreign words, uncommon phrases, names of journals or books, should be in italics. Italics may also be used to emphasize certain words or phrases in a scientific text.

For all chemical terms, the guidelines of the International Union of Pure and Applied Chemistry (IUPAC) should be used.

Punctuation Marks Inside and Outside a Quotation

This is a common dilemma encountered by scientists and engineers. The general guideline is to use the punctuation mark outside the quotation mark if only part of a sentence is the quotation, or if the punctuation mark is solely related to the sentence (not the quotation).

For example: To quote Professor Jacob, 'The expansion of the platinum under this condition is counter-intuitive. It goes to prove the unique nature of platinum'. In contrast, the following sentence, which is entirely the quote, will have the punctuation mark inside the closing quotation mark: 'The expansion of the platinum under this condition is counter-intuitive. It goes to prove the unique nature of platinum.'

Numbers

Numbers less than ten should be spelled out. They should be written as numerals when they are greater than ten. For example: 'six elements', '125 instances'. They should also be written as numerals when some unit of measurement (like 5 mm, 2 gm, 6 p.m.) follows the number.

When one list includes numbers both over and fewer than ten, all numbers in the list may be written as numerals; for example, 17 analog, 13 digital, and 2 radio chips.

All numbers at the *beginning* of sentences must be completely expanded. A sentence must never start with a numeral. For example: '15 products were released last year' is an incorrect way to start a sentence.

Large numbers, such as those in millions, should be expanded, but the number of millions may be written as numerals. For example: 20 million.

Commas must be used in numbers with four or more digits, after every three digits from the right. For example: 2,500,340. This makes it easy to correlate to thousands, millions, and billions. When referring to Indian currency, however, the comma is used to indicate crores, lakhs, thousands. For example: Rs. 14,46,82,000.

To describe ranges in pages, the least number of digits should be used (except in ranges in the group 10–19 in each hundred where the penultimate digit should be included). For example: 46–8, 3256–69, but 10–14, 1216–19. The penultimate digit is also used when referring to a range of years, for example, 2006–10.

Hedging

Hedging should be avoided. Hedging refers to the uncertain use of expressions such as maybe, perhaps, probably, suggest, etc.

Units

Units are abbreviated when they follow numerals. Otherwise, they are spelled out. For example: 5 mg, but five milligrams.

Standard Error of the Mean

Standard error of the mean (SEM) should be used in place of standard deviation (SD). This is because SEM reports the precision of an estimate of the mean in relation to its unknown value, while SD measures the distribution of individual results around an observed mean.

The ± Sign

This sign must be avoided as some journals discourage its use. If an observed mean has a notation such as 12.3 ± 0.4, it does not provide any indication whether the second figure is really a standard deviation or something else. This can be written more clearly by stating the mean first and the SD in parenthesis. Hence, a result may indicate a value by stating that the mean (SD) was determined to be 12.3 (0.4).

Equations, Chemical Reactions, and Formulae

All equations, chemical reactions, and formulae should be typed on separate lines unless they are brief, simple, and parenthetic in meaning. They should not run into the text. However, the equation remains a part of the text for purposes of punctuation. Short equations are centred; longer ones are aligned at the left margin and continued (if necessary) on subsequent lines. The larger subsections of the equation are broken just prior to each operator sign and aligned (with the operator signs). A set of equations should always be aligned to the equal-to (=) sign.

For example:

$$\underline{\hspace{3cm}} = (\underline{\hspace{3cm}})$$
$$+ (\underline{\hspace{3cm}})$$
$$- (\underline{\hspace{3cm}}) \qquad (2.5)$$
$$\underline{\hspace{3cm}} = (\underline{\hspace{3cm}})$$
$$* (\underline{\hspace{3cm}}) \qquad (2.6)$$

There should be sufficient white space above and below the equation, to separate it from the rest of the text. If equations or formulae are numbered for subsequent reference, the numbers must be placed on the right margin and must be enclosed in parentheses. There must be sufficient space between them and the end of the equation to avoid confusion. When referring to an equation, the term 'equation' must be used with the parenthetical number. For example: Equation 2.5. It could be in abbreviated form as Eqn. 2.5. Similarly, Fig. 3.2 can be used in abbreviated form to refer to a figure number.

Equation modifiers or defining statements should be left-aligned with the main equation and should not be centred. For example, 'for valency ≤ 4' or 'where $x = 5z/w$'.

SI units and Symbols

The System International d' Units (SI) is the form of the metric system advocated by the Canadian Standards Association (CSA).

The SI units and symbols constitute an essential part of the language of science. Metre, kilogram, second, ampere, kelvin, mole, and candela are the seven base units of SI.

The following are some of the guidelines for using SI units and symbols:

- SI units or their abbreviations should be used. The commonly used ones and their abbreviations are: metre (m), kilogram (kg), second (s), ampere (A), kelvin (K), candela (cd) and mole (mol).
- When the names of SI units are spelled out, the initial letter of the name (with the exception of Celsius) is not capitalized, except at the beginning of a sentence.
- A derived unit formed by division has 'per' between the units. For example: kilometre per hour, not kilometre/hour.
- A symbol represents a unit name and is the same in all languages.
- The symbols do not change in plural. For example: 10 kilometres = 10 km.
- The symbols are never followed by a period except at the end of a sentence.
- The symbol of a derived unit formed by division may be shown by using an oblique line (/) between the symbols in the numerator and those in the denominator (50 kg/m^2) or by the use of symbols with negative exponents (50 kg.m^{-2}). For example: km/h, not kmph or k.p.h.
- A space must be left between the numerals and the first letter of the symbol. For example: 320 lm, not 320 lm (to indicate 320 lumens).
- Sentences should not begin with symbols. As with numbers, the name should be spelled out.

Algorithms

A new algorithm should be described accurately and lucidly. It should give the reader an opportunity, if necessary, to compare the algorithm with other algorithms. An algorithm, for example, approximately optimizes some measure of performance. This performance measure should be explicit, especially in a new algorithm. It is always better to define the space of hypotheses that the algorithm searches when optimizing its performance measure.

When a new algorithm is introduced in a paper, experiments should be conducted to compare it with the state-of-the-art, not only for the same problem, but similar ones. The performance could be compared with a native standard like random guessing to add clarity. The performance criteria should be properly defined and explained.

All the limitations of the algorithm should be clearly described. A study of the cases where an algorithm is ineffective may help clarify the range of its applicability.

Negative Results

Negative results must be reported—they are important results, after all. A negative result is not required for the current investigation and may be due to an incorrect hypothesis. But this result may be of immense value to others. The negative result may prove to be a critical pathfinder in some unexplored area of scientific or engineering research.

4.2 Communication of Scientific or Engineering Results to a Broader Technical Community

A significant part of communication by engineers and scientists relates to the dissemination of results from experiments conducted. The following guidelines should be followed for this kind of communication.

- The experiment, if it throws new light on a topic, should directly justify its relevance and importance to the field of study.
- Measures of uncertainty stated at the conclusion of the experiment, should be in the form of estimates of standard error.

- To make the duplication of the experiments by other scientists easier, details about software and data should be included.
- The conclusions drawn from the experiments should be illustrated through a graphical display of experimental data.
- If the communication introduces a new algorithm, experiments should be conducted to compare it with the state-of-the-art, focusing on the same or similar problems. This would enable other scientists to compare the new algorithm with existing ones.
- Any limitation in the algorithm should be clearly stated. This would make the range of applicability of the algorithm clear to the readers.
- The experiment must be described with as much quantitative information as possible. This includes the processes, results, and analysis. If software is used, details of the software programme and data should be provided.

In particular, it must be noted that the experiments conducted must be based on the following:

(a) System model
(b) Numerical Analysis and Simulation

The two models helpful for detailing experiments, have been discussed below.

System Model

Assumptions are necessary in any experiment and are used to build a proposed model. Here, the model-assumption combination has to be stated clearly and unambiguously. These assumptions must make sense. Assumptions play an important role in any technical study and are necessary to make the problem mathematically tractable. However, it has to be ensured that these assumptions are the replica of some real world situations. If the proposed assumptions do not hold in general, then at least there should be some cases where these will hold. Figures should be used to explain this underlying model.

Numerical Analysis and Simulation

Numerical results are important and are generated as they are based on the model. These results have to be presented in such a way that the reader quickly and clearly grasps their significance. It is better to present them in the form of figures or tables. The parameter values must also make sense. These should preferably be chosen from systems in the actual world or at least should correspond to those values for which published results are available.

These results should be in the same form as existing ones. All results are to be carefully interpreted. Sometimes it becomes necessary to use simulation to validate the system model with assumptions. The technical report should include not only the average values of the simulation results, but also the statistical confidence intervals. Details should be provided about simulation time, the computer, and the language used.

An example of the system model and numerical analysis and simulation is provided in the following section, excerpted from a paper by C.P. Ravikumar and Rahul Kumar (2002):

Linear Regression Model
For each library, the least square estimation technique was used for arriving at a linear model of leakage power as a function of the number of cells. The linear model, described succinctly as, $y=mx+c$, has two important parameters: the slope m and the intercept c. These parameters are data-dependent, i.e., if the data points are scattered then the percent error in the prediction will be large. The slope and intercept, denoted here by S^{lib} and C^{lib} are shown in Table 5 for the three libraries. The leakage power model obtained is given by Eqn (2). Here, Cells is the number of cells in the design.

$$\ln P_{leak}^{lib} = S^{lib} \ln \text{Cells} + C^{lib}$$

...The above equation can be written as

$$P_{leak}^{lib} = \chi^{lib} \cdot \text{Cells}^{S^{lib}} \qquad (3)$$

The model has two parameters, namely S^{lib} and C^{lib}, which can be estimated using the following equations.

$$S^{\text{lib}} = \frac{\left(\sum \ln P_{\text{leak}}^{\text{lib}} \ln \text{Cells}\right) - \left(\sum \ln P_{\text{leak}}^{\text{lib}} \sum \ln \text{Cells}\right) / n}{\sum \left(\ln P_{\text{leak}}^{\text{lib}}\right)^2 - \left(\sum \ln P_{\text{leak}}^{\text{lib}}\right)^2 / n}$$

$$C^{\text{lib}} = 1/n \left(\sum \ln \text{Cells} - \ln \text{Cells} \cdot \sum \ln P_{\text{leak}}^{\text{lib}}\right)$$

Figures 4.1, 4.2, and 4.3 show the leakage power vs the number of cells for all the libraries and their corresponding linear models. Interestingly, we see that the parameter S^{lib} is close to unity. In other words, the leakage power dissipation grows approximately linearly with the number of gates. The percentage error in the predicted model for the three libraries is shown in Figure 4.4. The error was minimum for the slow library and maximum for the fast library. The plot in Figure 4.4 also shows the

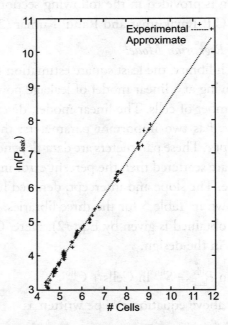

Fig. 4.1: $\ln (P_{\text{leak}})$ vs $\ln (\# \text{Cells})$ for Slow Library

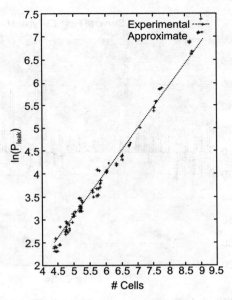

Fig. 4.2: ln (P_{leak}) vs ln (# Cells) for Fast Library

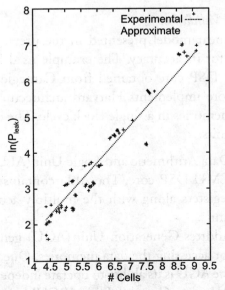

Fig. 4.3: ln (P_{leak}) vs ln (# Cells) for Fast

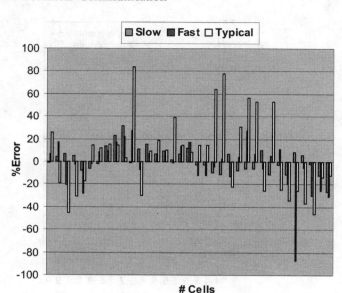

Fig. 4.4: Percentage error Vs (# Cells) for the Three Libraries

difference in the actual leakage power of the design and that predicted by the linear model.

Model Validation

The linear model presented in the previous section was validated for its accuracy. The example used for this purpose is a 24-bit DSP core obtained from Carnegie Mellon University. The core implements Harvard architecture and can read two data memories in a single clock cycle. The core architecture has four units:

- Data Arithmetic and Logic Unit (ALU) is the heart of the CMU DSP core. The ALU contains the X, Y, A, and B registers along with the multiply-accumulate and adder units.
- Address Generation Unit (AGU) generates the addresses for accessing the data memories. One important aspect of the AGU is its ability to operate independently of the ALU.
- Program Control Unit (PCU) contains the program counter (PC) and the flag bits for controlling the operation of the CMU DSP.

Table 4.1: Leakage Power Comparisons for the CMU DSP

| Component | # Cells | Leakage Power (mW) | | % error |
		Experimental	Linear Model	
ALU	7548	1.92	1.98	3.12
AGU	6500	1.59	1.71	8.22
PCU	3436	0.87	0.92	5.74
Bus Switch	458	0.16	0.14	-12.50
Core	17962	4.53	4.58	1.10

- Bus Switch generates instructions to move data among the memories. It works in conjunction with the address bus to accomplish the data transfer.

The core was synthesized by targeting TSMC's 0.18 micron standard cell library. The leakage powers obtained from the Synopsys Design Compiler for different units of the core and for the entire core are shown in Table 4.1. The leakage powers predicted by the linear model are also shown in Table 4.1. As can be seen, the estimator errors are on the positive side in most cases; for the case of the Bus Switch unit, the error is on the negative side, i.e., the estimator was too optimistic. In fact, the largest percentage error in prediction is also for the Bus Switch unit. We also see that the leakage power estimates for smaller circuits are less accurate as compared to those for larger designs. Thus, the model can be practical for predicting leakage power well in advance for large designs.

Here, the authors have used a linear regression model as the system model and have simulated it post-synthesis. The assumptions in this model are stated clearly. The results from the numerical analysis are very well presented using graphs and tables. The results also validate the model that has been used.

Finally, before experimental results are communicated, they must be checked to ensure that they are consistent with any expected or stated format.

The conventions concerning the formatting of experimental results that should be followed, are listed next.

A Few Formatting Conventions while Communicating Experiment Results

Typing

The text should be typed with double spacing throughout. There should be a 3 cm left margin and 2 cm right margin. There should be 2 cm margins at the top and bottom. This provides space for readers' or reviewers' comments.

Tables

Tables should have captions and column headings. The caption should be at the head of the table to enable readers to understand the table independently without consulting the text.

If the data can be given in the body of the text itself, then it is better to avoid tables or figures. Tables not referred to in the text are unnecessary.

Figures

Every figure should have a legend below to enable readers to interpret it without the help of the text. Each axis should be provided with a short informative title. All graphs, pictures, diagrams, and drawings must be termed figure. Figures and tables must be numbered in order of appearance.

Technical Clarity

Clarity is essential while communicating to a wider audience. With global communication becoming increasingly common, there is a critical need to adopt standardized technical conventions across the world. This makes generic communication simpler to understand and experimental results easy to compare against a global yardstick.

Appendix C provides a set of specific technical guidelines required by the IEEE, the ACM, and VDAT for publication.

5

Effective Use of Charts, Graphs, and Tables

Beautiful charts and graphs
etch a mark in the mind
and assist in the understanding
in a pleasant way.

SCIENTIFIC and engineering communication depends to a large extent on graphs, charts, and tables to explain and clarify the points stated in the text. Graphics also help to draw and retain the attention of readers.

Graphics must add value to a text and should not be included without good reason. These should not be attached at the end of a technical document, but should be interspersed on the same page for better understanding.

General Guidelines for Graphics

Deciding upon where to place graphics can be confusing. For best results, the writer should keep in mind the list given below.

- Keep graphics simple
- Place graphics vertically
- Place graphics near the relevant text
- Clearly indicate the source of the graphics, if borrowed

- Ensure the correctness of the data that is captured in the graphics
- Graphics should fit within normal margins. If necessary, enlarge or reduce the copies. There should be two blank lines above and below graphics
- Use the graphic tool that best communicates the theme
- Label the graphic form to ensure the theme is captured
- Use colour very rarely and only where it serves to amplify the message

Different Types of Graphics

Graphics serve different purposes and are of different types. The various kinds of graphics are illustrated below.

5.1 Bar Chart

This is the most commonly used method for the representation of data. It is used to compare values across categories. For example, the bar chart below clearly shows how the revenues, costs, and profits correlate in each quarter of a particular year. Thus, the message that profit as a percentage of the revenue has risen progressively from the 1st quarter through the 4th is clearly brought out by Fig. 5.1.

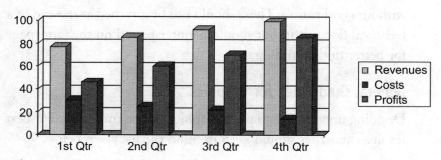

Fig. 5.1: Bar Chart showing Annual Rise and Fall.

Care should be taken to choose the right graph and co-ordinates that best amplify the desired message. For example, instead of correlating across revenues-costs-profits by quarter (as shown here), if the objective of the graph is to show how the revenues have grown by quarter, how the costs have ramped down by quarter, and the degree of increase in profits, then the following graph communicates the theme better.

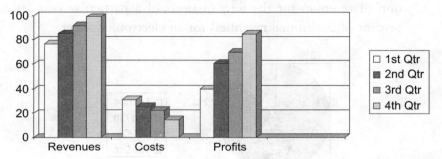

Fig. 5.2: Bar Chart showing Growth and Fall by Quarter

5.2 Line Chart

Line charts are ideal for capturing trends over a period of time. For example, in the line chart given below, it is easy to understand the trend of growth (or decline) in profits, revenues, and units shipped, as well as cost trends of a particular electronic gadget over the four quarters of one year.

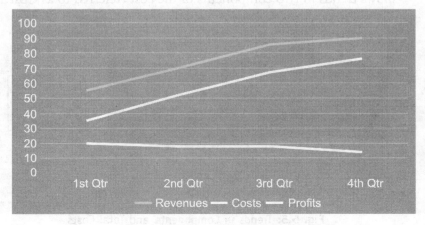

Fig. 5.3: Trends of Growth and Decline

5.3 Pie Chart

Pie charts are an effective way to depict the contribution of each segment to the total. There should be less than seven slices in a pie chart to enhance clarity. For example, Fig. 5.4 shows the proportion of revenues for the four quarters of a particular year (as a percent of the annual revenues) for an electronic gadget.

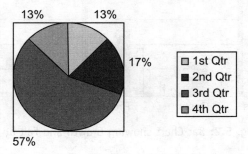

Fig. 5.4: Percentage of Revenues earned over Four Quarters

5.4 Area Chart

These charts display the trend of the contribution of each value over time or over categories. These are especially useful when depicting additive components. For example, the following chart shows trends in two components of the cost (referred to as Cost 1

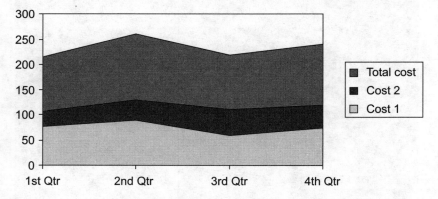

Fig. 5.5: Trends in Components' and Total Costs

and Cost 2) as well as the trend in the total cost (Cost 1 + Cost 2). Here, the areas for both Cost 1 and Cost 2 begin from the *X*-axis. However, a more intuitive way to describe an area chart is to show the cumulative areas of the components on top of each other. Hence, the top curve always shows the cumulative cost, and each of the 'individual areas' below show the value and trend of the components.

5.5 Radar Chart

This chart has a radial form, with markers at each data point. A study of the distances from the centre provides an intuitive way to compare data values across categories, as well as within a category. For example, in Fig. 5.6, it is easy to compare the profits, costs, and revenues in each quarter, as well as across quarters.

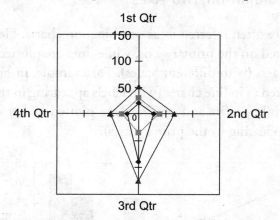

Fig. 5.6: Quarterly Growth and Fall.

5.6 Cylinder Chart

This is very similar to a bar chart, but has columns with a cylindrical shape.

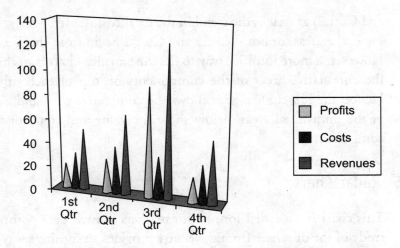

Fig. 5.7: Cylinder Chart showing Growth and Fall by Quarter

5.7 Line-column on Two Axes

These are often referred to as combination charts. Here, columns are plotted on the primary axis, while lines are plotted on the secondary axis (with different scales). For example, in Fig. 5.8, costs are plotted as in line charts (with labels appearing in the secondary axis), while profits and revenues are plotted as bar charts (with labels appearing in the primary axis).

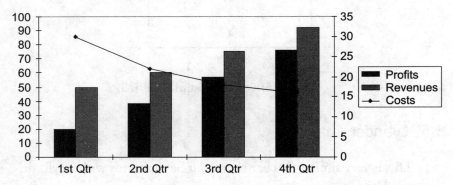

Fig. 5.8: Combination Chart of Line and Bar Charts

5.8 Column Bars

These are very similar to bar charts where values are compared across categories. The only difference is that the values in each category are depicted as horizontal columns, rather than as vertical bars.

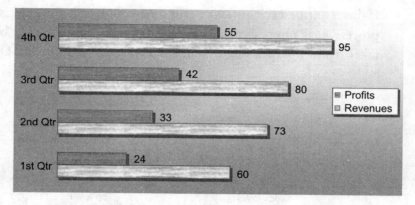

Fig. 5.9: Column Bars showing Growth in Profits and Revenues

5.9 Bubble Chart

In this chart, the size of the bubble depicts a particular aspect, while the X or Y-axis represents another. For example, in Fig. 5.10, the people count in the hardware, marketing, software, and

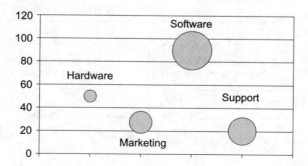

Fig. 5.10: Bubble Graph showing the People and Patent Count

support groups are depicted by the size of the bubbles, while the centre-point of the bubble along the Y-axis represents the number of scientific patents filed by each of these groups.

5.10 Flow Diagram

If the result of a scientific paper is a new flow or a novel decision process, it is often shown as a flow diagram. Fig. 5.11 has been taken from a paper titled 'Leakage Power Estimation for Deep

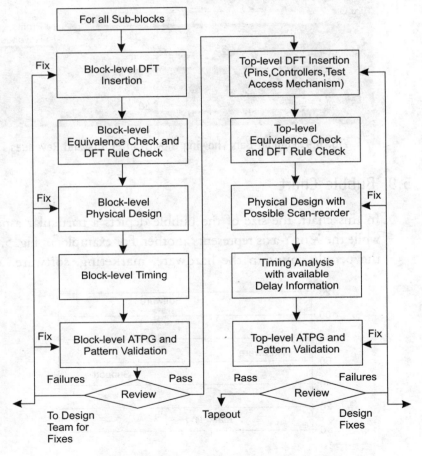

Fig. 5.11: Flow Diagram showing a Novel Process

Submicron Circuits in an ASIC Design Environment' by Rahul Kumar and C.P. Ravikumar.

5.11 Screen Capture

Often results are shown as a screen-shot. This allows the reader to get a glimpse of the computer output following a certain procedure. An example of a screen capture is illustrated below. This is excerpted from a paper, by C.P. Ravikumar and G. Hetherington, titled 'A Holistic Parallel and Hierarchical Approach towards Design-For-Test'.

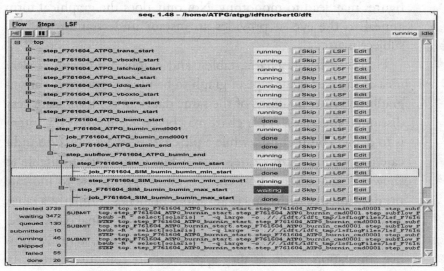

Fig. 5.12: A Screen Capture showing a Procedure

5.12 Tables

Tables are the most common method of depicting experimental results. At the top of each column is a column header, while at the beginning of each row is the row header. The main function of the table is to allow a rapid comparison of information. Data arranged chronologically in a table can indicate trends and patterns of rising or falling activity. An example, taken from a paper by C.P. Ravikumar and G. Hetherington, has been given below.

Design	Size (M gates)	Classical Approach	Framework
Design B	~ 10	Est. 30 days	7 days
Design S	~ 4.8	Est 30 days	6 days
Design WL	~ 2	Est 15 days	3 days
Design TE	~ 1	Est 10 days	2 days
Design J	~ 8	Est 40 days	6 days

Fig. 5.13: A table showing Various Designs and their Details

To conclude, graphs, charts, and tables are of great importance to scientists and engineers. The specific objective of graphics is to clarify the points made in the text. An assemblage of dry facts and figures may lead to monotony. Not only does the graphical form appeal to the eye, it provides an instant understanding of the subject matter.

However, before charts, tables, and graphs are drawn, a clear idea of the particular type of graph to be used is essential. Various graphical representations of the same data can communicate very dissimilar messages. A bar chart may be excellent when the intent is to depict comparisons, while a pie chart is better suited to indicate proportions across categories. A line chart is very suitable to show a trend, while a flow-chart is more appropriate to indicate the decision tree with choices at various branches. Similarly, a table, which is a very useful form to depict two-dimensional data, representation may fall woefully short if the intent is to actually show live simulation results as they appear on the computer screen. In the latter case, a screen shot is the best way to communicate the real-time experimental results. Ad hoc usage of the graphical form may not only be irritating to the reader, but it may actually convey a very different meaning.

Thus, a writer must first be clear about the message to be conveyed and choose a graphic accordingly. Colour should only be used in graphics if it adds clarity to the intended message. Also, too many charts, graphs, and tables may confuse readers.

6

Writing Technical Reports

A report is a report—
that provides information;
that helps decisions
on crucial matters.

A technical report is a statement of facts. It is a special form of business or technical communication and its purpose is to provide authentic information about a particular subject. It communicates new facts or presents new analysis, providing a fresh insight on a particular subject.

A technical report has a certain format that must be adhered to. It is not a sequence of text written in an unstructured manner. The purpose of a technical report has to be stated at the very beginning. This objective can vary depending on the subject of the report.

6.1 Objectives of a Technical Report

Technical reports may deal with different matters. These could be one of the following:

- root cause analysis of a particular failure
- evaluation of a new product or solution
- technical specification of a new system to be launched
- in-depth assessment of a unique situation

Each of these objectives has a common underlying foundation. It is based on an in-depth study of the topic and extensive causal analysis. It is critical to have the objective stated in a crisp, unambiguous manner.

6.2 Types of Reports

There are different types of reports for different purposes. The most common are:

- Technical evaluation report
- Technical specifications report
- Technical information report
- Research findings report
- Technical recommendations report

Technical Evaluation Report

The purpose of a technical evaluation report is to present necessary technical evaluation information in a logical way. This report is usually chartered by management to an expert technical team. The technical team carries out a thorough and in-depth evaluation of the subject. The team often interviews multiple internal and external stakeholders, conducts experiments, does extensive benchmarking, etc. as part of the evaluation.

The report helps the management to decide whether a particular course of action is feasible. It might be anything from installing a new gadget (which is a technological innovation) to assessing a new product roadmap (which requires a blend of technical and business evaluation). Technical evaluation reports are sometimes referred to as feasibility reports.

Some examples of technical evaluation reports are:

- evaluation of a new computer-aided design tool (which must include various benchmark data on its performance in real-life test cases)

- evaluation of the suitability of a new material in a manufacturing process (which must include information about how the material performs in normal conditions as well as in conditions of extreme temperature and stress)
- evaluation of an automotive product roadmap from various technical considerations (engine management system evolution, dashboard automation, etc.)

Technical Specifications Report

A technical specifications report provides details about the features of a product. It specifies how a product will work from a user's perspective. It also provides a clear insight into how a product works under various conditions. Technical specifications reports contain a large number of graphs and tables. Block diagrams of the system and sub-system are a key component of these specifications.

Technical specifications reports also provide interface information. It describes how a product interfaces with other products and systems. The information must be detailed to ensure that users have no difficulty or confusion during the product or solution integration process. Detailed technical specifications should also be an integral part of any patent disclosure.

Some examples of technical specifications reports are:

- technical specifications for consumer appliances (such as washing machines, televisions, cellular phones, etc.)
- technical specifications for components (such as aircraft or automotive engines, integrated circuits, connectors, etc.)
- technical specifications for software (such as how the software is intended to work in a system, various interfaces, etc.)

Technical Information Report

The main objective of a simple technical information report is to explain a technical subject clearly. This information allows readers to understand and appreciate system details. This report, however, does not provide any final recommendation—unlike a technical evaluation report.

Examples of such reports are:

- reports on the internal workings of an ultrasound system
- reports explaining the different video formats and standards
- reports detailing the complex reaction of two chemical substances

Research Findings Report

A research findings report summarizes the outcome of extensive research. It is often used to gain and share insight on a subject on which significant work has been done. Such a report includes a summary of past work, details of experiments conducted, and new approaches. Finally, such a report provides an account of all results obtained through illustrations and benchmark data.

Examples of such reports are:

- reports summarizing the most recent research findings related to behaviour of a new dielectric material
- reports with findings from research on new manufacturing methods to lower packaging costs
- reports that provide the findings of a research on improvement in the efficiency of the internal combustion process

Technical Recommendation Report

Technical recommendation reports provide very *specific* recommendations to those associated with the decision making process. These reports are a combination of both evaluation and recommendation and are immensely useful.

Some examples of such reports are:

- reports summarizing a product failure situation and recommending steps to alleviate future issues
- reports recommending structural modifications needed to handle overload

- reports summarizing different courses of actions that need to be taken depending on test results

6.3 Steps in Writing a Technical Report

The key steps to be followed while writing a technical report are:

- establishing clarity of purpose
- finding out about the audience
- collecting information and data
- analyzing the data collected
- structuring the report

Each of these steps is described in detail in the subsequent sections.

Establishing Clarity of Purpose

The fundamental purpose of a formal technical report is to convey information. The writer must plan the report carefully. To do this, she/he must first answer two questions:

a) What is the report about?
b) Who is it being written for?

The introduction section of the report should clearly answer these questions and should succinctly state the purpose of the report. People genuinely interested in and concerned with the stated purpose, will immediately appreciate the value of the report. To make this clear, most reports have a *written purpose statement* that provides a clear understanding of the objective of the report.

For example, if the purpose of the topic under discussion is to assess the reason for the field returns of mobile phones due to extremely fast battery drainage, then the written purpose statement would be as follows:

'Failure analysis of rapid battery discharge in mobile phones leading to field returns.'

The written purpose statement should be very specific. It cannot be vague or general. No irrelevant points or discussion should be included in the statement of purpose of the report. Even if there is significant amount of information collected, only the most relevant and the most important should be included.

For example, the following topics should be considered part of the report on 'Failure analysis of rapid battery discharge in mobile phones leading to field returns':

- number of phones in the field and factory likely to have been affected
- speed of battery discharge
- conditions under which the battery drains—especially stating if it is unique to a particular model or consignment
- methods to slow down the discharge
- possible reasons for the discharge and analysis of each reason
- summary of the root-cause based on experimental results and analysis

Finding Out About the Audience

Once there is clarity regarding the purpose of the report, the next task it to know the audience. It is critical to know who will read the report to ensure that it is written keeping the target audience in mind. The task becomes easy if the leader of the group or the company itself charters a team to develop and present the report. In that case, the report writer has a clear idea of what is expected and who will go through the report. However, it is possible that the report may be forwarded to many others if it is of significance to more than one group or activity.

Readers or audiences may be classified under the following heads:

Primary audiences: These are the people at the top, the key persons who take decisions on the basis of the report.

Secondary audiences: These are the people who will be affected by the action taken by the people at the top on the basis of the report.

Others audiences: These are the people who are interested in the report but are not directly involved.

The author of the report needs to clearly keep her/his audience in mind while preparing the report. The reaction of each of the above audience categories will differ. The author of the report should focus on the constituency that she/he cares most deeply for.

Collecting Information and Data

The step that is most critical in a technical report is collecting adequate and authentic information. A project stands on the credibility of the collected information. It is imperative to know where the relevant information is available. However, it is not enough to just collect vast amounts of data. The data and information collected must be carefully sorted to identify and root out 'impurities' or incorrect information.

Incorrect information is the bane of any good technical report. There are several reasons because of which the quality of data may become 'polluted'. The most common of these is the lack of credibility of some sources of data. As a result, there is a great possibility of the quality being inadvertently compromised. According to a business executive, 'The single-most challenging aspect for companies is to recognize and determine the severity of their data quality issues, and face the problem head-on to obtain a resolution. Spending money, time and resources to collect massive volumes of data, without ensuring its quality are futile and only leads to disappointment' (*Express Computer* 14/11/2005). The only solution to this problem is to carefully check the data at the very source before using it. It is essential for all the data to pass through a cleansing process first. In particular, information which is likely to form the basis of the recommendation, must be given importance.

A writer must be extremely cautious and choosy while collecting information from different sources. While analyzing the information collected, importance must to be given to the source of information. There should be no personal bias or preference.

Table 6.1: Sources and Types of Information

Type of Information	Sources
Primary	Observations, experiments, surveys, personal investigations, questionnaires
Secondary	Newspapers, journals, pamphlets, books, Internet search engines

As demonstrated in Table 6.1, there are two main sources of information: primary and secondary.

It is important to arrange the collected information in a methodical way. Careful docketing might prove useful. A particular type of information kept in a particular docket would facilitate (at a later stage) the effective sieving of the information collected.

Analyzing the Data Collected

Once high quality data is collected, it needs to be properly analysed. This is necessary to ensure that the information is really relevant to the purpose for which it is being collected. The report should be capable of helping authorities to take the appropriate decision. The report must also maintain objectivity. There is no room for personal bias.

Structuring the Report

A technical report should follow a logical structure. This structure allows the gradual unfolding of a story, which develops step by step till it reaches its conclusion.

Before the author begins writing the technical report, she/he must organize her/his thoughts and structure the report appropriately. The author must then put the material collected in proper order, eliminate unnecessary information, and select important ideas as part of the structure. Following this, it is always advisable to chalk out an outline of the report and use it to arrange the sections before beginning to write.

The two models used for writing technical reports are:

(a) The standard model
(b) The terms of reference model

The Standard Model

This is the most commonly used model for writing technical reports. The main feature of this model is that beginning with the introduction, it unfolds the storyline gradually and methodically. Everything in the report is supported by authentic data. Graphs and diagrams are also used extensively. The structure of the report under this model is provided below. It is important to follow this structure, though minor modifications are acceptable.

Report structure of the standard model

- Introduction
- Background and context
- Methods, methodology, or procedures
- Results and recommendations
- Conclusion
- References
- Acknowledgement

A description of each of these items has been provided below:

Introduction   The main function of the introduction is to provide an indication of the topic or subject of the report. This section clearly illustrates the purpose of the report. It describes the specific problem that has been investigated and the methods adopted to solve these problems.

The introduction section of the technical report should have an opening sentence such as:

The primary purpose of this report is to present an analysis of the call-dropping phenomenon seen in mobile phones when handoff occurs from one cellular base-station to the other. Based on many benchmarks and interoperability tests, the report gives a recommendation on how handset vendors and basestation players need to tune their Power Management systems.

Background and Context   This section provides a clear idea about the background that prompted the report. It provides a picture of the prevailing conditions which necessitated a thorough investigation of the situation. The section also suggests the scope of the solution-space—along with any constraints that must be adhered to in the final recommendation. This is different from the 'Recommendations' section, which goes beyond the solution space and provides specific and actionable solutions.

The background and context section could be as follows:

In the last 100 days, our company has received a large number of field returns and customer complaints regarding our mobile handsets. Our initial analysis shows that this happens under the following specific conditions:

- handoff speeds of greater than 60 kilometres/hour
- with the following handset models: AXA4300, B136429, MSD3458
- with the following base-station models: IEC1294, URT2485
- at nominal temperatures

The minimum expectation of our customers is for our company to have a solution that does not drop the calls abruptly during a conversation. A gradual fading, though not desirable, is a situation that our customers have agreed to accept. This report is intended to provide recommendations keeping the above customer expectation in mind.

Methods, Methodology, or Procedures   This section provides a comprehensive description of the procedure or methods adopted to arrive at the results. It clearly enumerates how the data was collected and the methods used to authenticate the data sources. It also states the specific problems that had to be overcome and the manner in which this was done. These facts are demonstrated with the help of appropriate tables, charts, and graphs. In fact, the validity or authenticity of the full report rests on this process.

The opening sentence of this section could be:

> Over 1,000 instances of failure were observed, with multiple operator and handset manufacturer combinations. Transmit and receive signals were measured in each case using a state-of-the-art interoperability equipment, such as EFR1296, coupled with the signaling software in benchmarks 1 through 15.

Results and Recommendations   The purpose of this section of the technical report is to present the key results and give recommendations to the sponsoring person or organization that asked for the report. Here the actual findings, secured after thorough study and investigations, are reported.

To make the results transparent and credible, they should be supported by appropriate tables and graphs. This is followed by the recommendations that are the outcome of the study conducted.

In some cases, the report might suggest further investigation. It may be noted here that the specificity of the recommendation is very critical to the reader. General guidelines or comments are not acceptable.

The results section of the technical report could read this way:

> A thorough analysis of the handoff failure instances has clearly revealed a problem in the Power Management circuitry in three models. These models are RTD1298, ERM4578, and TRT2385. We believe there are millions of units of these models in the field today and a recall of these will cause significant loss of company goodwill. Our failure analysis results on EFR1296 interoperability equipment on benchmarks 1 through 15, however, open up a new option. Dynamically increasing the voltage on the basestation equipment after it detects the above models leads to a slight loss of signal power and consequent fading—but avoids a dropped-call. This option will bring back our customer satisfaction, be cost-neutral to the company, and allow us to not lose our

goodwill. We strongly recommend making this change in the system board before end of this month.

Conclusion   A conclusion just concludes giving an overall view of the study and the key findings of the technical report. The intrinsic value of the study undertaken should be clearly stated here. There is no scope of stating anything new in the conclusion. It is always better to begin the discussion with a short re-statement of the most important points from the results section stating clearly what was actually done. In this connection, the significance of the actual findings can be clearly stated.

Acknowledgment   The 'references' section lists articles that the author of the technical report has consulted or referred to in the text. All publication details related to the references must be included in this section.

The purpose of the 'acknowledgement' section is to give credit and to thank those who assisted in the preparation of the report.

The Terms of Reference Model

The technical reports that adhere to the terms of reference model are slightly different from those that adhere to the standard model because, they are organized in a different way. The terms of reference format is used in routine reports. For example, a technical report that goes into the cause of corrosion of a high density metal, or the failure analysis of a chip under a unique combination of voltage, temperature, and pressure.

A routine report aims at collecting information and suggesting solutions. The scope of this type of report is smaller as compared to that of the standard model. Routine reports deal with relatively simpler issues. This is because detailed instructions concerning what is to be done, also referred to as the 'terms of reference', are stated at the beginning of the report.

When an author is chartered to write such a report, he/she must ensure that the norms of the 'terms of reference' are not violated. The report must be written within that framework. The frameworks containing the terms of reference may vary slightly in different kind of reports, but are inherently similar.

Some of the frameworks that can be used for such reports are:

Framework A

1. Terms of reference
2. Survey of the present system
3. Investigation
4. Results
5. Suggested remedy

Framework B

1. Terms of reference
2. Previous findings
3. Present investigation
4. Facts
5. Conclusions
6. Recommendations

Framework C

1. Terms of reference
2. Present method
3. Comment on the present method
4. Comparison with other methods
5. Suggested new method

Each framework begins with the terms of reference. The entire structure of a technical report is built on this foundation. These terms of reference are specific instructions that are given to the team or individual chartered to research, write, and present this report. These instructions clearly state the following:

- the particular type of information required
- the parameters within which the investigation should be confined
- the date by when the report needs to be submitted

The guidelines for the terms of reference of a particular report could be:

Major manufacturing issues arising from the intense heating of the chassis were recently encountered in the production line of our new X500 line of automobiles. This has caused significant yield loss in our factory in the last quarter. To alleviate this, we had a meeting with our large customers and our metal suppliers for the chassis. The management of these companies proposed that we introduce a new step in our manufacturing flow that treats the steel at a temperature of 185° Celsius prior to beginning the condensation process. Before this step is introduced across all our manufacturing lines in the country, the properties of the steel used in this process and its reaction with the Aluminum based alloy used in the prior step of the manufacturing process, need to be carefully studied.

Prepare and submit a report on the feasibility and impact of introducing this additional step in our manufacturing process. In the report, please point out the chemical reactions and related reliability side-effects that might occur between these two substances at this extreme temperature. Also assess all financial and personnel aspects that will be associated with implementing this proposal. These include:

- delta cost of introducing this step in our manufacturing process
- cycle time to be incurred in this annealing process
- availability of material science experts in our X500 team

The report is to be presented to the Group CEO within 15 days from the date of this memo.

The charter clearly indicates that this technical report should fall under the terms of reference model and must adhere to Framework A. It should include a survey of the present system, investigation, results, and suggested remedy.

Depending on the nature of the charter, it might sometimes be necessary to follow a certain framework of the terms of reference model. However, the standard model is more popular and is widely used as it adapts in a simpler manner to a linear flow of step-by-step analysis and recommendation.

However, irrespective of the structure of the report, the criterion of a good technical report is its accuracy. While presenting the information, a technical report should be objective and accurate. Information provided in such a report is extensively used by management while taking a decision. Inaccurate or flawed information or reasoning could derail the process. The accuracy of information depends on the sanctity of the data used. It is imperative, therefore, to ensure that the source of the data is authentic and reliable.

6.4 Guidelines for Writing a Technical Report

The quality of any technical report can be greatly enhanced by adhering to ten fundamental guidelines. Following the suggestions given below will add clarity and value to a technical report.

1. have a clear idea of the type of report to be written
2. clearly state the objective of the report
3. plan the sections and sub-sections
4. write the report headings in bold letters
5. make the contents specific and remove ambiguity
6. ensure that the contents meet the needs of the readers
7. use diagrams, flowcharts, and graphs to enhance content/results clarity
8. use as many examples and illustrations as needed to drive home the message
9. avoid jargon of all types
10. use a proper layout to draw attention to key information

The key guidelines concerning the structure of technical report have been listed next.

- The title should be descriptive of the problem, but should be crisp and clear.
- There should be a list of appropriate keywords. Keywords help readers to identify the actual domain of the report.
- The introduction of each technical report should clearly state the thesis of the paper. It should also provide an organizational plan for the gradual and logical development of this thesis.
- The main body of a technical report contains relevant data and is the most important part of the report. It should be clear and informative. It must state succinctly what was done to arrive at the recommended solution. Wherever necessary, graphs and diagrams should be used to illustrate the points. The focus in this section should be on the technical content of the paper. It should include the relevant points that would help authorities come to a decision easily.
- Assistance of any kind received from any source should be properly acknowledged.

To conclude, a technical report should be scientifically organized and should provide a clear and informative description of the problem investigated. The report should follow a logical structure. The natural unfolding of the storyline should enable a reader to clearly and easily grasp the subject. To achieve this, it is necessary to go from the generic to the specific. Also, it must be ensured that the data, information, and detailed analysis directly support the recommendation(s), as the findings of the report help decision makers to make the right call. Finally, the last criterion of a good technical report is that it must be written in a simple language.

7

Precis of Science and Engineering Related Topics

A précis is a concise version of a big thing;
telling in a few simple words
what the big one says in many words.

THE ability to write a good précis is essential for a scientist or an engineer. Many technical conferences and journals ask authors to reduce their original submissions to a third of their original size for the final camera-ready copy. Scientists and engineers, after preparing and sending their original submission in a detailed form to enable expert peer review, often find themselves scrambling for ways to reduce the size significantly. This poses a challenge, as they do not want to lose any critical content in the process. Often such page limits are not explicitly stated, but are expected from the author. It is important in such cases, to re-assess the original text for brevity.

Authors could face a similar situation while writing technical reports. Their detailed data-collection, investigation, and analysis often lead to a draft report, which is typically voluminous. Before the report is submitted, it needs to be reduced significantly to provide greater clarity, which will simplify the decision-making process.

A writer may face the same constraint while sending an e-mail to her/his peers. She/he may have to provide an abridged version of a detailed product requirements document or a summary of a voluminous product specification.

In all these situations, what is needed is a systematic process that takes a large document and converts it into one that is significantly smaller, while retaining critical content.

7.1 What is a Précis?

A précis is a clear and concise abstract of a passage that retains its correct and full intent. It is a replica of the original work, but one that is trimmed to the minimum without losing critical content. The new contents are put down in such a way that it reads like the original composition and gives a clear idea of the subject matter of the original passage. A précis is not a paraphrase. It lucidly expresses the main ideas of the author in a concise manner. It discards all unimportant points and retains only the main ideas of the original text.

7.2 Techniques of Writing a Précis

While writing a précis, the most important task is to ensure that the original sequence of events and the flow of ideas remain unchanged. These should not be tampered with in any way. The main objective of the writer of a précis should be to eliminate all unimportant points, discard high-sounding or multi-syllable words, and trim the sentences. If the original passage has been written by someone other than the author of the précis, it must be ensured that the précis does not contain words (other than keywords) from the original passage or document. The précis should not contain the reflections or the personal views of the précis writer. Views not expressed in the original passage have no place in a précis. It should be a fresh, crisp document which is more readable than the original.

7.3 Guidelines for Writing a Précis

Usually the original document and the précis are authored by the same scientist or engineer. A writer often has to re-write a scientific paper or technical report to meet submission guidelines. However, there may be instances where a writer may have to send to a peer or a manager a summary of the product specifications (from an extensive document and datasheet written by the marketing or applications team) to enable them to write a precis.

The guidelines for writing a précis are:

a) Read the original text (scientific paper, technical report, product document, etc.) carefully and try to understand it well.

b) Read the original text again to glean out the main theme.

c) Read the original text once again to pinpoint the main ideas of the original text.

d) Underline the main points (this will also help eliminate unimportant points).

e) Read the original text once again to make sure that no important points have been missed.

f) Note the main theme and the important points.

g) Preserve the sequence of the points.

h) Read all the notes that have been made in prior steps carefully.

i) Without consulting the notes, try to write the substance of each of these points in your own words.

j) Check to make sure that all important points have been covered.

k) Make a draft of the précis on the basis of your notes.

l) Read what you have written to make sure that the text flows smoothly from one point to another.

m) Count the number of words in the original passage and then count the number of words used by you.

n) If the number of words in the précis is more than one third of the number of words in the original text, eliminate some words by condensing your précis. Reduce the number of words to one third (or to the stated limit).

The ability to fully comprehend the original text is essential for the précis writer. The précis writer must understand the meaning of the original text before attempting to reduce it. The précis writer should also possess a good command over the English language so that the ideas contained in the passage can be expressed lucidly— without replicating the words from the original passage.

7.4 A Few Significant 'Don'ts' in Précis Writing

Here are some important traps that need to be avoided when writing a précis.

- Never be verbose.
- Never introduce a new idea or opinion which is not part of the original text.
- Never use abbreviations or contractions.
- Never begin with expressions such as: 'the paragraph is about', etc.
- Never repeat a point.
- Never deviate from the mood or tone of the original text.
- Never use figures of speech, such as similes, metaphors, etc., which may be part of the original text. Also avoid polysyllabic or high-sounding expressions.
- Never modify the views or ideas of the original text.
- Never use introductory remarks such as: 'I want to draw the attention of the readers to....' , 'The objective of this discourse is to find ....'
- Never include casual remarks or unnecessary details (unless these are pertinent and necessary for a proper understanding of the text).
- Never put unnecessary emphasis on a point unless it has been emphasized in the original text.

The logical order of the ideas presented in the original paper or document should never be disturbed. These must appear in the same order in the précis as in the original text. In other words, the continuity or the flow of ideas, as found in the original text, must be retained. Just picking up some words or expressions from the beginning, middle, or end of the original text without any semblance of coherence does not make a précis.

It is always preferable to draw up a first draft of a précis, go through it carefully, and then write the final précis. The précis, when completed, should never look like the précis or summary of another document. The completed précis should retain the appearance of the original.

Finally, it is always better to check the clarity of a document or report after it is written. This can easily be done by means of a simple test known as the clarity index or fog index. A detailed account of the fog index is provided in the first chapter of this book. Of course, if the above instructions are followed meticulously, the document will have high clarity.

The précis writer has to condense the ideas expressed in the original paper, report, or document. He/she has to structure the new text, which is as important as the content. This needs to be done in such a way that the same ideas can be expressed using fewer words.

For example, if a sentence in the original is, 'It has been found that for proper maintenance of the machinery, servicing and proper cleaning should be done every year', it could be re-written as: 'Annual servicing of the machinery is required'.

The proper usage of words and the correct method of writing sentences for scientific purposes have been dealt with in Chapters 1 and 2 of this book. In conjunction with these, the principles of writing a précis as stated above should be followed.

Here are a few examples that of précis writing:

Example 1: This example is generic in nature and may form the basis of correspondence between professionals discussing the state of the IT industry in India. The difference in clarity between the original text and the précis is apparent.

Original Text:

The Information Technology (IT) industry has witnessed a remarkable growth since the turn of the century. It has changed the perception of India around the world and brought the 'made in India' brand to the forefront. Talent management and retention has arisen as a major issue in the far-changing work scenario in the IT field. But the downside to all the progress has been an increasingly unmotivated worker, who is ready to change jobs at the drop of a hat. The mercenary behaviour the IT industry is also responsible for high attrition rates.

In today's competitive marketplace, employees are well informed and know their exact market value and the opportunities available to them. Earlier, people who were not performing optimally, were informally mentored, however, the multinational manager today has little patience with low performers.

IT has brought India rapid recognition, but the balance could as easily be disturbed if we are not on guard. The unreal salaries are bringing on a negative net effect on the industry. There is a grave danger for the country because talented professionals are being picked up from other industries leaving a lacuna in crucial sectors of engineering and manufacturing.

An important motivating factor for staying on in a job is when one has a circle of friends with whom one shares common interests, whether it is community service, hiking, or Hindi music. Close friends at work can be a crucial reason for not quitting, even when the job situation is stressful at times. As in the military, 'country' may be an awe-inspiring motivating factor, but the core in the regiment is the ultimate uniting strength. So also in an organizational context, the company provides the overarching culture, but what blends the employees is a close-knit team.

The salient points of the text are:

- There is a remarkable growth in the IT industry today.
- With this has increased the attrition rate.
- Employees are now aware of their market value.
- As a result of attrition, big gaps are being created in important sectors.
- A common bond amongst the employees may be a binding factor.

While the main objective of writing a précis is to significantly reduce the text, special care should be taken to ensure that no important point is omitted.

Here is the précis of the original text based on the notes taken earlier:

Précis:

The Indian IT industry has witnessed remarkable growth. However, this has brought with it an associated problem. Employees, now conscious of their market value, are looking for better opportunities. This has resulted in high attrition. The demand for talented professionals has also created big gaps in important sectors. The only way to hold employees is to find a common bond, which will bind them.

Example 2: This example depicts a text used in the scientific community on the origin of weights and measures.

Original Text:

We often hear that science is a revolutionary force that imposes radical new ideas on human history. But science also emerges from within human history, reshaping ordinary actions, some so habitual that we hardly notice them. Measurement is one of our most ordinary actions. We speak its language whenever we exchange precise information or trade objects with exactitude. This very ubiquity, however, makes measurement invisible. So it is not surprising that we take measurement for granted and consider it banal. Yet the use society makes of its measures expresses its sense of fair dealing. That is why the balance scale is a widespread symbol of justice.

The men who created the metric system understood this. They were the prominent scientific thinkers of the Enlightenment, an age which had elevated reason to the 'sole despot of science'. These savants, as the investigators who studied nature were known in those days, had a modern face looking toward our own times, and an older, glancing back toward the past.

They were appalled by the diversity of weights and measures they saw all around them. Measures in the eighteenth century not only differed from nation to nation, but within nations as well. This diversity obstructed communication and commerce, and hindered the rational administration of the state. It also made it difficult to compare results with colleagues internationally.

A précis of the above text, written using the steps stated earlier, is:

Précis:

The impact of science, though forceful, is invisible. This is evident from the case of weights and measures. The creators of the metric system, who were the savants of the Enlightenment age, were shocked by the different weights and measures circulating in different countries. Naturally, this diversity caused a hindrance in the process of communication and commerce. This also created difficulties in comparing the results in weights and measures with other nations.

Example 3: This text captures the use of IT in the medical world.

Original Text:

Telemedicine, digital X-rays, pacemakers working with multiple microprocessors, and information technology have brought healthcare a long way forward. Right from the process of admitting a patient to a hospital, through the diagnosis and all the way to treatment itself, IT is perhaps the most significant aspect after, obviously, the doctor.

Information technology is perhaps at the very core of healthcare delivery. Today, quality healthcare revolves around hardcore diagnosis and diagnostic equipment like scans are driven by IT. On the whole, it is a technology-driven approach. The advances of IT in the medical field have helped improve efficiency and accuracy as well as speed up decision-making. There are some very good software that can tell the doctor that two drugs are not combining or they do not get along together. There are also many types of software that will tell you the tentative diagnosis if you feed in all the symptoms.

Moreover, telemedicine too has advanced by leaps and bounds over the years. Over 300 hospitals across the country are now linked to the telemedicine network using different networks and satellites. Not only can a doctor view X-ray and scan reports of patient sitting miles away but he can also scan the patient and even assist other doctors in surgeries.

Here is a précis of the above text:

Précis:

Information Technology (IT) plays a vital role in the healthcare system today. Due to it, sophisticated diagnostic equipment like digital X-ray, pacemaker, scanning equipment, etc. are available. There are various kinds of software that not only help detect the compatibility of two drugs on a patient but also help quick diagnosis. Because of the great advancement of telemedicine, hundreds of hospitals are now linked to the telemedicine network. Now a doctor can treat patients sitting miles away.

Example 4: This example shows the convergence of engineering and biology, along with its impact.

Original Text:

Employment of biomedical engineers is expected to create a boon through 2012. The aging of the population and the focus on the health issues will increase the demand for better medical devices and equipment designed by biomedical engineers. The demand is growing from hospitals, healthcare companies, medical electronics companies, and career options from abroad. Earlier hospitals did not hire in-house biomedical engineers. But with the realization that in-house biomedical engineers can take immediate action as and when required, hospitals big and small now prefer employing in-house biomedical engineers. Biomedical Computers, an essential part of biomedical engineering, has certain specializations.

Bioinstrumentation is the application of electronics and measurement techniques to develop devices used in diagnosis and treatment of disease. Computers are an essential part of instrumentation.

Biomechanics includes the study of motions, material deformation, flow within the body and in devices and transport of chemical media constituents across the biological and synthetic media and membranes. Progress in this field has led to the development of artificial heart and heart valves, artificial joint replacements as well as better understanding of the function of the heart, lung, blood vessels, capillaries, etc.

Clinical engineering involves the application of the latest technology to healthcare. Clinical engineers are responsible for developing and maintaining computer databases of medical instrumentation and equipment records and for the purchase and use of sophisticated medical instruments. They may also work with the physicians to adapt instrumentation to the special needs of the physician and the hospital.

Here is a précis of the above text:

Précis:

Biomedical engineering is expected to flourish in the near future. Various factors such as the ageing population, attention to health issues, and medical equipment designed by biomedical engineers are the key contributing factors. The demand for in-house biomedical engineers is also growing. Biomedical computers are an essential part of biomedical engineering. It has specializations such as Bioinstrumentation, Biomechanics, and Clinical Engineering.

Example 5: This example is a discussion on the critical issue of mobile number portability.

Original Text:

Imagine this: you are frustrated with your mobile service and want to dump it for another operator. However, you have given this number to hundreds of clients over the last few years. It is an unnerving task to inform all of them of the new number besides getting the new number printed on new visiting cards, letter heads, et al. Assume, now, that you are allowed to retain your mobile number even after you have jumped from one service provider to another. The single word that can make this possible is 'number portability'.

All across the globe, number portability has been introduced by regulators of their countries, to bring in competition and encourage new players—leading to lowering of tariffs for customers. Number portability allows customers not only to move from one mobile service provider to another within the Global System for Mobile Communication (GSM) to Code Division Multiple Access (CDMA) services (you have to exchange your phone in this case) and also from landline to wireless phones.

Sound exciting? The hitch is that Indian telco operators are not enthused to the least. And for a change, both GSM as well as CDMA mobile operators are united on this new issue—they think the time for introducing number portability has not yet come.

Here is a précis of the above text:

Précis:

Mobile number portability would be of great advantage to a subscriber dissatisfied with an existing mobile service provider. The problem today is that the phone number changes with the change in the mobile service provider. Only number portability can solve this problem. This has been implemented in other countries where the service provider can be changed without disturbing the existing number. But the problem remains unsolved in India, as service providers are not agreeable to this idea.

To conclude, being acquainted with the techniques of writing a précis would be of great help to a scientist or engineer. The executive going through a recommendation, the reader of a technical report, or the recipient of an e-mail would appreciate it if these documents came in an abridged form. A précis written in the prescribed form, following the correct techniques can go a long way in ensuring effective communication.

8

Speech Communication: Effective Presentation Techniques

When you are communicating,
you are dealing with people;
to know how to handle them
you must understand them.

EFFECTIVE communication is an essential requirement of leadership. Scientists and engineers, especially those holding key positions or possessing critical technical knowledge, are often required to communicate their views. The main concern of the speaker is to influence the thinking and subsequently the actions of those who listen to her/his speech. This ability to effectively communicate and persuade is of great importance to a scientist or an engineer. Success depends on this ability to penetrate the mind of the audience with the speaker's power of persuasion.

Communication through speech requires a psychological insight into the human mind. The art of communication depends on the skilful application of this insight. While communicating, one is dealing with human beings. It is essential to have a clear idea of how normal people behave and act under diverse conditions. The objective of the speaker should be to mould or shape the thinking process of the listeners to enable them to grasp the presenter's point of view. Hence, the first task of the speaker is to be aware of this thinking process of the audience.

8.1 Elements of a Good Speech

Delivering a good speech requires taking the audience into confidence. The audience comes with the expectation of learning something of value. If they trust the speaker and have confidence in her/him, they will automatically imbibe the ideas put forward by the speaker. Reciprocally, the speaker must play her/his part by being honest, sincere, and articulate.

The speaker has to imagine that she/he is taking the audience through a journey. It is a journey from an unknown to a known world where the aspirations of those listening to the speech could be fulfilled. When the speech begins, the audience has an open mind, with no idea about what the speaker is going to say. A forceful and resourceful speaker helps the audience understand the purpose of the speech gradually. The viewpoint of the speaker must be absolutely clear. There should be no lack of transparency. Only then will the audience unhesitatingly follow the lead provided by the speaker.

A well-prepared speech can leave its mark on the audience. Such a speech, in addition to good preparation, requires excellent delivery. This implies successful communication of the essential ideas of the speaker to the audience. To achieve this, the speaker has to:

- focus on the message
- speak with conviction and enthusiasm
- use her/his voice forcefully and effectively

The speaker must bear in mind that the purpose of communication is to convey a message and not just to talk. Communication, in fact, is a two-way process. During this process, the speaker must seek feedback from the audience. The gestures and postures of the audience automatically reveal their reaction. Bright, attentive eyes focused on the speaker reflect the response of the audience. It automatically establishes a rapport between the audience and the speaker. This prompts the speaker to proceed, keeping in view the following immediate goals:

- The speaker must win the confidence of the listeners and acquaint them with something about which they may have only a vague idea.
- The speaker has to convince the listeners, reasonably and logically, that the proposals being mooted will benefit them. This message should be short, clear, and capable of being easily understood by the audience. This message can be given in a single sentence or better still, in a short phrase, so that it remains engraved in the minds of the audience.

The main elements of a good speech are:

- organizing the speech (also referred to as 'developing the storyline')
- gauging the mood of the audience
- skilfully using the art of audience persuasion
- using voice as a powerful tool
  - speed
  - articulation
  - pause
- effectively using body language
  - hands
  - feet
  - eyes
- leveraging visual aids
  - front projectors
  - flip charts
  - video and audio tapes

These elements are described in the following sections.

8.2 Organizing the Speech (Developing the Storyline)

The most crucial task of the scientist or engineer is to carefully organize her/his own thoughts, and subsequently, the points that need to be presented. These must be arranged in a simple and logical manner. It is best to begin with a storyline. The storyline clearly and succinctly takes the audience through the message that the speaker intends to communicate. It gives a clear idea of the overall theme. This theme then needs to be placed in different *compartments*. The storyline, thus, acts as a roadmap that guides the audience from one compartment—a significant point of the presentation—to the other. This mechanism engages the audience, as it takes them through this virtual journey.

Listening to the presentation, the audience first becomes curious, then interested, and finally glued. The people in the audience then begin reacting mentally to the points being made by the speaker. At the end of a good speech they are convinced, though gradually, about the argument presented by the speaker. They align with the speaker's point of view. This is how the speaker's objective is fulfilled.

As is evident from this sequence, structured organization of a speech is essential to ensure effective communication. Random thoughts interpolated anywhere take away from the objective of the presentation. Speech organization has another positive advantage—it is of great help to the speaker. This is because the logically arranged train of thought keeps the different ideas tightly linked—with one idea following the other in a systematic pattern. This helps the speaker to remember and recapitulate her/his thoughts easily.

Speakers may be 'positive' or 'negative'. Both positive and negative speakers use different techniques to organize their speeches. A positive speaker, in the psychologist's terminology, organizes her/his thoughts in a way that motivates the audience. The speaker leaves the audience with no doubt that what she/he is expostulating will serve their interest. As most people think of themselves first, this approach has an immense advantage.

A negative speaker, on the other hand, is one who speaks of everything apart from that which directly serves the personal interest of the audience. This usually happens when the speaker does not organize her/his speech. Hence, it is vital for the speaker to plan and organize all thoughts, ideas, and recommendations, keeping in view the interest of the audience.

Closely linked to mentally sequencing the speech, is planning the modalities of delivery. A well-planned and sequentially organized speech allows the speaker to clearly articulate his/her thoughts. The combination of these two dimensions—of sequencing content and planning delivery—makes it easy for the audience to comprehend, remain engaged, and internalize the key message of the speaker. On the contrary, if listeners have to rack their brains to understand what is being said, it is likely that they will lose interest. Using short, simple words, with a well-planned and sequenced theme appeals to the audience.

Abraham Lincoln, in his famous Gettysburg address, used a remarkably simple technique: 267 words, of which 206 had only one syllable! The power of this speech also lies in strong organization. It begins by recalling with pride the responsibility conferred on the current generation by the fathers of the new nation. It then states the reason why the audience is present on that occasion. Finally, it concludes with a passionate call for action—as the best way to honour the memory of those who fought for the nation and made the highest sacrifices.

The Gettysburg address:

Four score and seven years ago, our fathers brought forth upon this continent a new nation: conceived in liberty, and dedicated to the proposition that all men are created equal.

Now we are engaged in a great civil war…testing whether that nation or any nation so conceived and so dedicated…can long endure. We are met on a great battlefield of that war.

We have come to dedicate a portion of that field as a final resting place for those who here gave their lives that the nation might live. It is altogether fitting and proper that we should do this.

But, in a larger sense, we cannot dedicate…we cannot consecrate…we cannot hallow this ground. The brave men, living and dead, who struggled here, have consecrated it, far above our poor power to add or detract. The world will little note, nor long remember, what we say here, but it can never forget what they did here.

It is for us the living, rather, to be dedicated here to the unfinished work which they who fought here have thus far so nobly advanced. It is rather for us to be here dedicated to the great task remaining before us...that from these honored dead we take increased devotion to that cause for which they gave the last full measure of devotion...that we here highly resolve that these dead shall not have died in vain...that this nation, under God, shall have a new birth of freedom...and that government of the people...by the people...for the people...shall not perish from the earth.

As is evident from the above speech, short and simple words used by the speaker, combined with a clear organization of ideas brought forth spontaneously and passionately, can move the audience and make a lasting impression.

8.3 Gauging the Mood of the Audience

Gauging the mood of the listeners is essential for effective communication through speech. The speaker can either continue with the structure according to her/his planned storyline, or modulate it while speaking. This is a decision that the speaker has to take, depending on the prevailing mood, feelings, or attitudes of the audience. A problem can crop up when the speaker, unmindful of the mood of the audience, makes the speech logical, methodical, and systematic and states one key point after the other. This is possible with scientists and engineers as they are used to structured thinking. They are not used to dynamically modulating their thoughts. However, any speech will inevitably fall flat if the speaker does not feel the pulse of the audience at the very beginning and at every stage of the presentation.

It is necessary, therefore, is to observe the reaction of the audience and act accordingly. Audience reaction can be gauged by looking closely at the audience to discover the focus of their attention. If their attention is fixed on the speaker, it clearly indicates that they find the speech interesting. If not, the speaker immediately needs to do something that will be of interest to the audience. She/he must change the style of delivery or mode of presentation in such a way that it will absorb the interest of the audience.

The Role of the Right and Left Hemispheres of the Brain

It is important to understand the role of the right and left hemispheres of the brain in communication. It has to be noted that effective presentation is a creative process that goes on in the human brain. There are different modalities and skills that back up this creative process. The right and left hemispheres of the human brain have independent functions and have unique and special attributes. Each of these, therefore, has an important role to play in the process of spoken communication.

The left side of the brain is the seat of reason—concerning logical and analytical thinking. The right side of the brain is the seat of emotions and feelings. It controls artistic and creative functions, and is more concerned with the heart. There is a link between the two halves of the brain. Messages are transmitted across a thick band of nerve fibres called the *corpus callosum*.

A close look at spoken and written communication will help illustrate the difference in the functioning of the right and left brain. For example, in any presentation there is flow or spontaneity. However, in reality, occasional rambling and lack of logical reasoning are often part of spoken language. This happens because in speech communication, the right brain has a fair play. On the other hand, written communication is governed primarily by the left brain and a methodical analysis of the situation, a logical approach, and syntactical accuracy are the foundational elements.

During speech communication, it is not the right brain that rules in all cases. Many speakers, especially scientists and engineers, rely significantly on their left brain for their presentations.

However, it is critical that speakers modulate their style (left brain versus right brain) in accordance with the mood of the audience.

This is best illustrated by a comparison of the speeches of Brutus and Mark Antony in Shakespeare's *Julius Caesar*. Brutus and Antony were Roman senators. Both eulogized Julius Caesar—but used very different techniques. Brutus failed to gauge the mood of the audience and focused on reasoning. His arguments, emanating from his left brain, were based on solid reason. He enunciated his reasons for taking Julius Caesar's life and asked the people to use their power of reasoning. He said with an emphatic gesture: 'censure me in your wisdom, and awake your senses, that you may the better judge.'

However, Brutus' fault was that he failed to gauge the mood of the audience. The terms—wisdom, reasoning, better judge—were at that time alien to the audience. Hence, he did not connect with the audience and the impression he made was fleeting.

Antony, in contrast, understood the audience much better. He leveraged his right brain and connected instantaneously with the audience. He was fully aware of mob psychology and utilized his power of carrying the people with him. When he first got there, he realized that he had to face an audience that had turned hostile due to Brutus' speech. He gauged their mood—and was prepared to counter it.

He carefully stated early in the speech: 'I come to bury Caesar, not to praise him.' After that, quite deliberately, he repeated four times: 'Brutus is an honourable man.' The first time the audience did not catch the irony or sarcasm in the statement. But thereafter, Antony, effectively using his right brain, went on to provide numerous illustrations to prove his point. He stated how good, benevolent, and unambitious Julius Caesar was. Instantly, the audience began to realize the undercurrent of irony and sarcasm that ran through his speech.

Antony realized that his aim had been achieved when one of the citizens remarked: 'If thou consider rightly of the matter, Caesar has had great wrong.' This reaction was precisely what Antony wanted. He achieved the objective of his speech. Brutus did not.

This example illustrates how critical it is for a speaker to connect with the audience by gauging their mood and modulating the content, length, and style of the speech accordingly.

8.4 Using the Art of Audience Persuasion

To achieve the objective of convincing the audience, the speaker should try to influence and appropriately modify the thinking process of the listeners. This task is not easy. It requires an understanding of the human mind, as well as the necessary skill to articulate effectively. The speaker must be acquainted with the subtle art of persuasion to motivate the audience.

The speaker must quickly state, in the first few minutes of the presentation, the benefits that the audience will derive by listening to the entire speech. These benefits must be stated, not from the speaker's perspective—but from the perspective of the listener. This could mean a *complete overhaul* of the message and a fresh look at the contents. While delivering the presentation, the speaker must keep a keen eye on the reaction of the audience and modulate the speech accordingly.

The Five Deadly Spoilers

Inability to persuade the audience can be the result of the five deadly spoilers, which have been explained next.

a) lack of confidence of the speaker
b) audience benefit doubtful or unclear
c) proposals suggested too complicated for the audience to fathom
d) speaker's negative approach that disheartens the audience
e) speaker's inability to enunciate points clearly

Of these, the first spoiler—lack of confidence—is the most widespread spoiler. To persuade the audience, the speaker must exude confidence and enthusiasm. This confidence and enthusiasm should be evident from what is said, how it is said, the tone of the voice, and even facial expressions, gestures, and posture. The speaker must be sincere, straightforward, and true to him/herself. As Shakespeare states in *Hamlet* (Act I, Scene II), 'This above all: to thine own self be true'. The speaker should not try to impress the audience with flamboyant jargon. What is said must come from the heart; only then will he/she be able to persuade and win the hearts of the audience.

Avoiding the next few spoilers is essential. A speech should clearly and succinctly communicate its benefits to the audience and influence them through its clarity and simplicity.

To illustrate this, the abridged version of a speech given by the leader of a technology company to a large gathering of televison manufacturers is provided below. He was communicating to them a revolutionary breakthrough in chip technology that would dramatically change the industry cost structure.

The Chip that Cheers

Have you ever seen a television set that has dimensions more like a movie screen? That has a display which is sharp and wide? That gives you a home theatre experience with CD-quality Dolby-digital surround sound? That takes viewers straight to the cricket stadium—or to the mountainous terrains of Tiger Hills? Yet…and yet costs you less than today's standard-definition television sets? And has an ability to dramatically transform your company's revenues and profitability?

Well, thanks to a breakthrough patented semiconductor technology that our company has recently sampled to select customers, we can achieve this. This is not something you may possibly have in the future. Yes! It is available TODAY! And as unbelievable as it may sound, our initial customers have reported a completely different order of clarity in the television sets that use our technology. They have seen benefits arising from the ability to display twenty times more pixels on their HDTV screens.

Let us now talk about the cost. You will assume that this HDTV set will cost you double the standard definition television set. The reality is, it is actually ten percent less expensive! This can really put the TV market in your target region completely on fire.

How did we achieve this? First, we came up with a completely new material in our chip design process. It cost us a tenth of the material cost that is commonly in use today. Of course, as you know, many other companies have tried to use this material in the past—but failed. They failed to get the right manufacturing yields while using this new material. BUT...we believe our company has solved this. I received the most current data just hours before I came here...and I am delighted to report to you that we have touched an unbelievable yield of 99.2%.

Next, we came up with a breakthrough innovation in architecting a low cost video processing chip engine. It achieves a processing throughput that is five times today's best processors. It supports all the standards that are in use today and is even compatible with the two key ones that we see evolving in the future.

But what does this mean to you—as manufacturers? This chip will let you lead the market when you build your TVs, with this novel technology. You can actually offer consumers an experience of tomorrow, with the cost that is lower than today's! It will result in significantly higher revenues and increased profitability for you. The chip that cheers will make you win!

An analysis of the speech indicates that it observes the criteria of a good and effective speech. To persuade the audience and capture their instant attention, it opens with the attention-grabbing title: 'The Chip that Cheers'. He starts his speech by clearly focusing on the benefits to the audience—increased revenue and profitability. He talks not from his company's perspective, but in a way that relates to the audience.

To eliminate any possibility of the audience remaining unconvinced, the speaker earns credibility by stating the positive experience that their early select customers have derived. He builds credibility further by providing some high-level details of the new chip.

Finally, he concludes on a positive note—stressing not on the consumer benefits but the real financial benefits (revenue and profitability) that the members of the audience will be able to earn for their companies. At this stage, there is a high probability that the audience has been successfully persuaded to procure his company's chips.

To achieve their objective of audience persuasion, speakers at times resort to another technique. They use a collection of words or phrases as *slogans* a number of times to drive home their point. Reputed Greek orators used this technique (called 'anaphora') effectively.

An example is a speech by Winston Churchill, who used this technique during World War II to motivate and enthuse his leaders. The speech was made when the city of London was being devastated by German bombers. An excerpt is provided below:

> …we shall go on to the end, we shall fight in France, we shall fight on the seas and oceans, we shall fight with growing confidence and growing strength in the air, we shall defend our Island, whatever the cost may be…

In this speech, Churchill used the expression 'we shall' a number of times which helped stir the hearts of the people.

The same approach was used by Martin Luther King Jr. He used the expression, 'I have a dream', a number of times in his historic Civil Rights speech.

> I have a dream that one day this nation will rise up and live out the true meaning of its creed. We hold these truths to be

self-evident: that all men are created equal. I have a dream that one day on the red hills of Georgia the sons of former slaves and the sons of former slave owners will be able to sit down together at a table of brotherhood. I have a dream that one day even the state of Mississippi, a desert state, sweltering with the heat of injustice and oppression, will be transformed into an oasis of freedom and justice. I have a dream that my four children will one day live in a nation where they will not be judged by the color of their skin but by the content of their character. I have a dream today.

These speeches illustrate how critical it is to skilfully and effectively use the art of persuasion. This helps the speaker to connect strongly with people and gain instant appeal.

8.5 Using Voice as a Powerful Tool

The tone and modulation of voice plays an important role in a presentation. It is always better to speak in one's natural tone without any artificiality. To avoid monotony, it is necessary to vary the tone depending on the situation. Just a change in the tone of the voice is capable of expressing the feelings of the speaker. By doing so, the speaker can emphasize a point, to which special attention must be drawn. Alternatively, the speaker can express abhorrence or displeasure just by modulating the tone of her/his voice.

The three main techniques used by speakers for highest impact are speed, articulation, and pause.

Speed

The pace of the speech should neither be too fast nor too slow. Every speaker has a natural pace. However, it has to be ensured that listeners face no difficulty in understanding what the speaker is saying. On the other hand, if the pace is too slow, the audience may feel bored. In short, the speaker's speed of speaking should be such that the audience can understand what is being said, without any conscious effort on their part.

Articulation

Articulation means properly pronouncing a word without slurring over syllables or just mumbling something incoherently. The inability to articulate is the product of careless colloquialism that is often resorted to during informal conversation. These have to be rigorously avoided in a formal presentation where every word must be pronounced clearly and distinctly. There should be no impediment between what the speaker is talking about, and the audience's proper understanding of it.

If the ideas of the speaker are expressed in a distinct and pleasing manner and in a convincing and confident tone, her/his confidence will be transmitted to the audience. The audience will then gradually feel confident that the proposals and actions suggested by the speaker, if executed, will be beneficial for everyone concerned. The art of clear and confident articulation, if cultivated well, will always help the speaker achieve her/his objectives.

Pause

A deliberate pause can be significant, and in fact, more eloquent than spoken words. It helps the audience understand the feelings of the speaker and what she/he wants to convey. A pause at the proper time and at the proper place serves another useful function: it gives the audience time to recapitulate and assimilate what the speaker has said so far. However, just as all things should be done in the right proportion, so also is the case with the pause. The motto here should be: 'no less and no more than just what is necessary.'

These then are the basics that prompted Abraham Lincoln's eloquent rhetoric. They also stimulated Swami Vivekananda's and Winston Churchill's inspiring oration. It is also the basis of Martin Luther King Jr's rousing speeches that spearheaded the civil rights movement.

Table 8.1: Body Language and What it Implies

Body Language	What it reflects
Standing erect	self-confident
Standing, but bent forward	frustrated and dejected
Sitting with legs stretched, apart	relaxed and at ease
Arms crossed on chest	defensive attitude
Hands touching the cheek	deeply contemplating
Rubbing the eyes	disbelief—is it true?

8.6 Body Language

It is possible for a person to express many things without saying a single word by means of body language. As the speaker stands before the audience, even before the speech starts, the audience is liable to judge or categorize her/him. Through the speaker's body language, the audience learns whether the speaker is confident, in complete command of the situation, and well-prepared. The speaker's face is an open book to the audience. It tells them if the speaker is nervous and strained, or relaxed and eager to speak. Body language, in fact, is a vital factor as it exposes the inner person—the nature/attitude of the person standing in front of the audience.

Examples of body language and what it may indicate have been shown in Table 8.1.

Distracting mannerisms communicate the wrong message—timidity, restlessness, and nervousness. Hence, the speaker should guard against inadvertently sending out such messages to the audience.

Here are some examples of distracting mannerisms that indicate awkwardness and lack of self-confidence:

- Clearing the throat frequently—this is a signal of hesitation and nervousness.
- Swallowing too often when the gaze of the audience is fixed on the speaker—this is an indication of a dry throat caused by the onset of nervousness.

- Frequently wetting the lips with tongue and swallowing repeatedly—this also indicates nervousness on the part of the speaker.

The speaker's gestures and posture are no less important. These are discussed below.

Hands

Hands can be of great help to a speaker and at the same time could be a formidable hindrance. They help the speaker to emphasize the points being discussed. As long as they are not over-used, hands help the audience understand more clearly the points being made. However, sometimes speakers fumble and do not know what to do with their hands while speaking. They change the position of their hands frequently. The speaker should keep her/his hands on both sides if they are not being used otherwise (to hold papers or a pointer).

Feet

Constantly shifting one's feet and aimless pacing also do not make a favourable impression. These are indications that the speaker is nervous. The reverse, standing still, is also equally awkward. The best course is to retain one's composure and behave normally, looking confidently at the audience.

Eyes

The best way to exude confidence is to maintain eye contact with the audience. The method is simple—look at the audience as often as possible. A good way is to look straight into the eyes of a member of the audience for a few seconds and then shift one's glance to another person. It is better to make eye contact with as many people as possible. After that, a general glance at the whole audience completes the contact. This will help everyone feel involved and a healthy rapport will be established between the speaker and the audience. This is an important factor in speech communication.

8.7 Visual Aids

We have come a long way from the days of powerful orators who depended entirely on their own power of articulation—without any support of technology. In contrast, speakers today have been greatly benefited by the advances in science and communication technology. Visual aids have become useful tools that help speakers deliver messages effectively.

These aids, if properly used, add impact to a speech by intensifying the interest of the audience. The main advantage of a visual aid is to provide notes/graphics while the speaker is engaged in presenting. This amplifies the speakers message.

Visual aids have three main functions:

a) The speaker is automatically reminded of what she/he is going to say next

b) They allow the speaker to move about freely with confidence

c) They allow the speaker to maintain direct eye-contact with the audience

While a speaker typically speaks about 100 words per minute before an audience, the audience, on an average, thinks about four hundred words per minute. There is a difference between the speed of speech and the speed of the thought process of the audience, which can distract the audience from the communication. Attractive visual aids partially alleviate this situation as the eyes of the audience remain glued to what is being shown, keeping them interested at all times.

Due of the advancement of science and technology, highly sophisticated business presentations are now possible with the help of computer assisted slides. While using these slides, a few points are worth remembering:

- Too much information should not be crammed onto a single slide. It is always better to spread out the information in different slides.
- It is necessary to be consistent in the presentation, especially when demonstrating columns of figures. If in the first illustration, the years are presented in the X-axis across, they should not be shown in the Y-axis in the next.
- Text should be used sparingly in the slides. Words should be used to only explain the *message* from the slide being shown.
- It is better not to overuse the laser pointer randomly over the slides as this may distract the audience.

The most important task for the presenter, however, is to select the appropriate visual aid for the presentation. Descriptions of visual aids used by presenters most often, have been provided below. The most commonly used visual aids are:

- front projectors
- flip charts
- video and audio tapes

Front Projectors

A front projector is commonly used in a conference room. It is a flexible aid and allows one to make graphical slides on the computer until just before the presentation and display these immediately on the screen. Like any visual aid, if properly used, it can enhance the quality of the presentation.

A few simple tips while using the front projector as a visual aid are given here.

- The screen should be placed so that the participants have a full view of it
- The screen should be at a level above the heads of the audience
- The view of the audience should never be impeded
- The room should be appropriately darkened

- The speaker should talk to the audience and not to the screen
- Laser pointers should be used sparingly and only to emphasize a point

The slides used in conjunction with the front projector and screen are commonly prepared using Microsoft PowerPoint. The rapid developments in the features of graphic slides has made them much easier to prepare.

Flip Charts

Flip charts are commonly used in relatively smaller settings, such as an interactive work-session. This is often followed by a presentation that refers to the contents noted in the flip charts.

Some key considerations while using a flip chart are:

- The height of the easel should be appropriately adjusted before the presentation.
- Each page should have a proper headline.
- Different colours should be used for page headings and also to highlight important points.
- Pastel colours should be avoided. Colours such as black, blue, etc. are more appropriate.
- It is advisable to avoid writing up to the bottom of the page as it might be inconvenient for those sitting in the last few rows.
- Sufficient time must be allowed for reading, retention, and note taking.
- Not more than ten lines should be written on a page.
- For pages with complex drawings, it is always better to prepare ahead in light pencil and then trace with a marker. Alternatively, these may be drawn visibly, ahead of the work-session.

Video and Audio Tapes

Video tapes can serve as a powerful visual aid during speech communication. They often serve as a refreshing departure from the monotone of live speech. Playing tapes transports the audience to a different world. Some key points to be taken into account while using video tapes before and during the presentation have been given below:

Some important tips to be used before the presentation are:

- The video tape should be rewound to the starting point for next use
- The audio levels and contrast/brightness should be checked
- The lights should be dimmed, but not turned off

Some important tips to be used during the presentation are:

- The purpose of the tape should be explained
- The audience should have a clear idea of what they are expected to learn from the tape

Guidlelines for Using Audio Tapes

In contrast to video tapes, audio tapes serve a somewhat different purpose. An audio tape is usually interspersed in the speech in short bursts as part of a presentation. For example, a specific quote from an expert can be played to the audience to emphasize the speaker's point. In scientific or engineering sessions on audio performance, playing an audio tape may also help drive home a point.

While playing the tape, the volume should be adjusted so that all participants can hear it clearly. Only a high quality recorder should be used to prevent distortion. The recorder should be placed at the table level of participants.

Finally, the purpose of the audio tape should be explained to the audience before it is played. The audience may be told to make notes of points they did not understand, which could be clarified by the speaker at the end of the session.

It is important to choose the audio-visual aid that is most appropriate for the session and the audience. It needs to conform to the subject matter and style of the presentation. However, these aids should never be used as a crutch. No attempt should be made to dazzle the audience with excessive use of audio-visual aids when these are not necessary.

Creating Rapport with the Audience

It is extremely crucial to understand the importance of establishing a rapport with the audience and the means that a speaker may apply to do so. Different speakers adopt different methods to establish this rapport. Story telling, if it supports and elucidates the viewpoint of the speaker, is one of these methods. Here is an example from the speech of a great savant, Swami Vivekananda.

In his historic address at the Parliament of Religions, Chicago, September 1893, Swami Vivekananda began his speech by addressing the audience as 'Sisters and Brothers of America'—which helped him establish an instantaneous rapport with the audience. After that, he gradually elaborated on his own thesis saying that there could never be any division—any rift—between man and man because of religion. He went on to connect further with the audience by saying that the basis of all religions is the same. Nobody could claim, 'my religion is superior to any other religion'. However, mankind was oblivious of this fact. He said, 'I am a Hindu. I am sitting in my own little well and thinking that the whole world is my little well. The Christian sits in his little well and thinks that the whole world is his well. The Mohammedan sits in his little well and thinks that is the whole world.'

To support his views, Swami Vivekananda told a story:

> A frog lived in a well. It had lived there for a long time. It was born there and brought up there...One day another frog that lived in the sea came and fell into that well, to which it asked, 'Where are you from?' The other frog said, 'I am from the sea'. 'The sea, how big is that? Is it as big as my well?'—and he leapt from one side of the well to the other. 'My friend', said the frog of the sea, 'how do you compare the sea with your little well!' The frog took another leap and asked, 'Is your sea so

big?' 'What nonsense you speak, to compare the sea with your well!' 'Well then', said the frog of the well, 'there can be nothing bigger than my well, the fellow is a liar, so turn him out'.

From his speech it is evident that an appropriate story told at an appropriate time can not only help to establish the views of the speaker, but also strengthens his/her rapport with the audience. Scientists and engineers could intersperse a story in their speech to reinforce their arguments.

Finally comes the concluding part of the speech. This is the place where the speaker has to stop. The speaker needs to be aware of when and where to stop. There is no point in repeating the same point over and over again, as the audience is likely to lose patience. To end the presentation, it is best to summarize the points discussed, as is normally done at the conclusion of a written paper. A fitting finale would be to end with an interesting remark, a quote, or an appropriate punch line. The audience must leave with a positive impression, with the feeling that they have been told something that enthuses them. As the saying goes:

Tell 'em what you are going to to tell 'em
Tell 'em
Tell 'em what you told 'em.

9

The Role of E-mail in the Communication Process

E-mail has brought the world closer;
it is a good master,
but a bad servant
if not used properly.

I N the field of communication, people are aggressively discarding the old order and welcoming the latest innovations. The key drivers of this transformation are increased speed and efficiency provided by these innovations. Electronic-mail (or simply, e-mail) is one such vector.

The importance of e-mail in the corporate sector and in academia is well established and the use of e-mails amongst professionals has increased rapidly. However, to be effective, a more scientific style of writing e-mails should be adopted. An e-mail must express crisply and unambiguously the writer's thoughts. An e-mail must also reflect the personality of the writer. This aspect is important as an e-mail is more personal than a scientific paper or a technical report.

This chapter is not intended to be a treatise on the mechanism of sending an e-mail. The details on e-mail software are easily available in manuals that come with the package. Hence, such preliminaries have been dealt with only in a nutshell. What is required for effective communication through e-mail is a thorough knowledge of the working details of the electronic network.

9.1 What is an E-mail?

An e-mail is the electronic equivalent of the ordinary letter. E-mailing, in simple words, is the act of sending and receiving a message through a computer. It is a very effective technology that allows one computer user to communicate with another specified computer user. The process is quite simple. Once written and sent, the e-mail is transferred to an e-mail server. This mail server forwards a copy of the message to the addresses of the recipients as indicated in the e-mail. The transfer protocol used for this is called Simple Mail Transfer Protocol (SMTP). According to this protocol, the e-mail server at the transmission end makes a connection with the e-mail server of the recipient and the message is transferred. Although different e-mail systems use different formats, there are some recent techniques that are making it possible for users of all systems to send and exchange messages in compatible formats.

Domain names in an e-mail address specify the location of the person's account on the Internet. They very often show the type of organization the sender belongs to. This is done using the following naming format that is used as the suffix.

com	Commercial organization
edu	Educational institution
org	Non-profit organization
net	Network
gov	Non-military government organization
mil	Military government organization
int	International organization

Sometimes the e-mail address also has a country suffix:

in	India
au	Australia

ca	Canada
fr	France
it	Italy
jp	Japan
uk	United Kingdom
us	United States

9.2 The Usefulness of E-mail

The usefulness of e-mail for scientists and engineers, both in the corporate world and in academia, is immense. E-mail is not only less expensive and faster, it also has an additional advantage. It can be transmitted anywhere, at any time, irrespective of the location and the time zone.

A great advantage of e-mail is that the recipient can seek clarification on any matter immediately and receive a response within minutes. Even when the communication needs are bulky and span multiple megabytes of information, e-mail allows the sending of such documents (for example, reports, notes, slides, sound files, graphics, etc.) easily as e-mail attachments. The recipient can respond with immediate feedback upon receiving these documents. Further, the same message can be transmitted to a number of recipients without additional expense. Likewise, the physical distance between the sender and the receiver has no impact on the cost.

The receiver of an e-mail is not interrupted during work when a particular message comes to the mailbox. She/he can handle e-mails according to her/his own time and convenience.

The facility of using e-mail off-line when required is another important and useful feature of electronic mail. This allows the sender of a message to write as many messages as needed, even when she/he is disconnected from the network. These can be saved as e-mail drafts and placed in a folder and can be sent when it is convenient for the sender.

It is also possible for the receiver to disconnect from the network after retrieving all e-mails from the mailbox. This reduces connection hours and consequent expenses. It also allows the recipient to browse through the mails when required and at a convenient time.

9.3 Limitations of E-mail

E-mail has a few limitations, but these are not insurmountable. It is possible to steer clear of most of these by paying a little attention. Some of the limitations are:

- It is possible to gauge the personality of the sender from an ordinary postal mail. However, in the case of an e-mail, it is difficult—though not impossible.
- Except inside firewalled settings where a message may be recalled, there is no scope for re-thinking the contents of the message after it is sent. Once a message is sent, there is no way to stop an on-line recipient from accessing it immediately.

Spam

A vexing menace for e-mail recipients is 'spam'. Spam is nothing but junk e-mail. Such unwanted e-mail clutters the inbox of the recepient. It is necessary to take appropriate steps to delete or divert this type of mail. Many organizations have specialized software to detect and block spam. However, these are never ironclad, and there is always the possibility of the inbox being cluttered with e-mails.

Hackers

One must also be prepared for a breach of privacy. While all e-mail systems are protected by passwords, e-mail hackers pose a serious threat. Also, sharing passwords or having simple passwords (such as, the same word as the user-name or the name of a family member) makes it easier for hackers to enter the protected system.

Differing Formats

Another problem for the sender of an e-mail is that what appears on the screen after composing the message might look different from what the recipient receives. The difference could be due to a change in text format (for example, a message created in 'plain or rich text' could automatically be converted to 'html') or change in screen dimensions. It could change significantly if the underlying structure (hardware and software) of the recipient's e-mail system is very different from that of the sender.

Distancing Factor

In addition, e-mail, unlike face-to-face conversations, video-conferencing, or even telephonic conversations, does not convey emotions or feelings well. This is because of the absence of vocal inflections and gestures that convey emotions while interacting in person. Thus, e-mail often becomes a dull and emotionless medium. Worse, subtle emotions could be misinterpreted by the recipient.

9.4 Emoticons: Emotional Icons Used in E-mails

To help the sender of an e-mail express emotions more effectively, 'emoticons' or emotional icons made their appearance. The main function of these emoticons is to lend some colour, diversity, and emotion to a message. The sender can express himself in three different ways:

- by using visual emoticons, such as a smile, as in :-). This rotates the face clockwise by 90°.
- by using character emoticons, such as <vbeg> (very big evil grin)
- by using short-hand expressions like PMFJI (pardon me for jumping in)

The difficulty with emoticons is that there are no standard defini-tions for some of them and so the intended meaning has to be guessed. Some examples of visual and character emoticons are provided below:

:-o	expresses surprise
>:-(	expresses anger
:-)	expresses happiness
:-(	expresses sadness
>:-(	expresses anger
:-l	expresses indifference
:@	expresses shock
:-/	expresses indecision

9.5 E-mail Shorthand

E-mail shorthand (or e-mail acronyms) consists of acronyms used by e-mail users that have evolved over a period of time. Here are a few examples:

IMO:	in my opinion
IMHO:	in my humble opinion
IAC:	in any case
TYVM:	thank you very much
BTW:	by the way
FYI:	for your information
NRN:	no reply necessary
TIA:	thanks in advance
TTYL:	talk to you later
IMCO:	in my considered opinion
IAC:	in any case
BBFN:	bye bye for now

9.6 Internet Jargons

Internet jargons are words taken from the English language, which have expanded in meaning over years of usage in the electronic medium. Here is a list of some Internet jargons:

Junk mail or spam: This is unwanted and unsolicited mail sent by an individual or by a company simultaneously to many people.

Ping: It is a device to ascertain if the recipient at the other end is ready or available.

Lurking: This means reading a message in a chat room or on-line forum, without revealing one's identity and without posting any comments.

Flame: This refers to an abusive and hostile message that may have far-reaching consequences.

Bot (from robot): This term refers to software that is capable of acting in place of or on behalf of a remote human being.

Mailbot: This refers to software that has the capability of automatically responding to an e-mail.

Bounce: When a particular message does not reach the sender, it is said to have bounced. This may be due to an incorrect e-mail address or some problem at the receiver's end.

9.7 Network Etiquette

An e-mail, though an informal method of communication, must conform to some basic rules. These rules are referred to as 'netiquette', the short form of network etiquette.

Some style tips are:

- All 'caps' should not be used. In an e-mail, it is equivalent to 'shouting'. This creates a bad image of the sender in the mind of the receiver.
- A good e-mail should have a proper structure. It becomes easier for the reader to read the message on the screen when it has a proper layout.
- Line length should be limited to 65–70 characters across. This avoids text wrap-around.
- The e-mail should be as short as feasible. Key points to be responded to or read by the recipient must be stated in the first five sentences. Long e-mails are often skipped by recipients.

- In the specified place for sender's address, the sender must identify her/his full name and not just the e-mail address. This helps the recipient identify the sender easily.
- While sending a reply to a mail, instead of clicking on 'new message', the 'reply' button must be clicked. This will ensure that the recipient will get the original message, in addition to the reply. It will serve as a quick reference and will save the recipient some time as she/he will not have to search for the original message.
- One needs to be careful when sending the same e-mail to a number of persons. The e-mail addresses of all these persons are put in the 'To' box. Thus, each recipient of the mail not only finds out who the other recipients of the e-mail are, but more importantly, their e-mail addresses as well. However, it is unethical to divulge the e-mail address of a person without prior permission. One way out is to put all addresses, other than that of the primary recipient, in the BCC field (blind carbon copy). The name of everyone in the CC (carbon copy) goes openly with the message contents, but the names of people in the BCC list cannot be seen by others.
- One should be conscious while using the 'To' field in an e-mail. When more than one person gets a message, there might be confusion about who should take action or reply to the message.
- One should not take the etiquette blunders of others seriously. If offended, it is best to send a message, such as, 'I couldn't understand what you actually meant'.
- It is important not to use expressions, such as 'immediate', 'urgent', 'important', unless absolutely necessary.
- There is absolutely no place for sarcasm in an e-mail. It will very likely lead to unpleasant consequences.
- Likewise, it is important not to 'flame'. In Internet jargon, to 'flame' someone means to write an e-mail using an abusive language.

- Sometimes an e-mail contains multiple points that are expected to be answered by the recipient. In such cases, it is best to number each point.
- Wherever possible, it is best to avoid attachments. It is more convenient for the recipient to go through the contents, if these are included in the body of the e-mail itself. There are several reasons for avoiding attachments. These are:
  - Attachments take time to download.
  - Attachments take up much space on the recipient's computer.
  - For people using small form-factor handheld devices, attachments are not always displayed properly.
  - Attachments could be carriers of viruses. A properly manipulated virus can have a devastating effect on the recipient's computer.
- When large attachments are sent, these should be compressed using standard compression utilities such as winzip. Prior permission of the recipient is necessary if large files are being attached.
- Any links provided should begin with http://. This will allow the recipient to click on the address and go directly to the site, without having to copy the URL.
- Subject lines are critical to attract the attention of the recipient, who receives hundreds of e-mails per day. Hence, what will interest the recipient truly must be stated here.

9.8 Computer Viruses

In the terminology of computer security, a virus is a 'self-replicating/self-reproducing automation program that spreads by inserting copies of itself into other executable code or documents' (Wikipedia 8/3/2006). Its behaviour is similar to a biological virus, which spreads by inserting itself into living cells. In fact, computer jargon associated with the word 'virus' is also taken from medical terms. This insertion of the virus into a particular programme is termed 'infection'. The infected file is called the 'host'.

Viruses could be extremely disruptive. A properly manipulated virus can have a devastating effect on the Internet worldwide due to its ability to destroy all types of data. According to an expert estimate, a worm called 'Mydoom' infected approximately 200,000–300,000 computers worldwide January 2004 (Times Online 8/3/2006). In March 1999, a virus called 'Melissa' forced Microsoft and a number of other large companies to shut down their e-mail systems until the virus was identified and eliminated. Similarly, in the year 2000, the 'I love you' virus caused devastation, though it was reported to be a mild virus.

According to a study by Symantec Corp titled 'Internet Security Threat Report', published during the first half of 2005, the number of new viruses targeting the Windows Operating System has increased by 48%. More shocking is the revelation that an increasing number of hackers try to disrupt the functioning of the web for financial gains—not just for the thrill, as was the case earlier. According to the study, the viruses exposing confidential information accounted for 74% of the 50 most damaging viruses, up from half earlier. (cxotoday 8/3/2006). Fortunately, however, with the introduction and growing popularity of powerful Internet anti-virus software, it has been possible to control viruses.

9.9   E-mail Signature

A unique feature of e-mails is e-mail signature. The signature acts as a letterhead for the sender and/or for the company. The signature effectively conveys the particulars of the sender—designation, contact information, etc. Most signatures include this information, referred to as 'business card' information, and also include a couple of lines about the sender of the mail. This information gets appended *automatically* at the end of each e-mail from the sender.

It is advisable to make the signature not more than 4–6 lines long and also not too wide. For example, a signature could look like this:

Ranjit Bose
Executive Director, Base Electronics
http://baseelectronicssoft.com
e-mail: ranjitbose@baseelectronicssoft.com
Phone: +91-80-25643587 (work)
+91-9812345678 (mobile)

9.10   E-mail Discussion Groups

Discussion groups on the Internet allow people who have similar interests, but live in different places, to communicate. A number of such discussion groups exist today in every organization as well as across the globe.

Discussion groups have a common e-mail address that is used to address all subscribers. A subscriber is likely to begin getting a very large number of e-mails almost immediately after signing up to such discussion groups. Hence, it is advisable to get information about the scope and other features of a discussion group before joining it.

9.11   Filtering the Flow of E-mails: Folder Management

Countering the in-flow of e-mails can at times become a problem for a busy executive or for any person belonging to a high volume mailing list. Professionals today receive numerous e-mails everyday. This is because all the mails sent are stored in an e-mail server. A lot of time could be lost in going through these mails and sorting out the urgent ones. One way to counter this problem is to install an e-mail software or script that sorts out and scientifically organizes the e-mails. This script is often referred to as a filter and works in conjunction with folder management. Folder management is a key feature of e-mail services. It allows the recipient to sort out incoming (and outgoing) messages into clearly defined folders.

Most e-mail programmes identify certain patterns in the message and sort it into the appropriate folder. Some of the e-mails received may be from large groups. The e-mail filter can identify e-mails coming from such addresses and can automatically direct them to a designated folder. These folders, storing e-mails from and to groups, are best named with the word 'list' to identify the source as a group mail. Likewise, all e-mails received from a particular engineering manager may be directed to another folder, with the name of the sender as the title. It may be noted that filters may also be applied to 'sent' e-mails. This allows the e-mail user to review all messages exchanged (sent and received) with key contacts. Spam and other e-mails that are known to be of little interest can be directed to a 'deleted' folder—and removed periodically.

Once these filters are set up, there are usually only a small number of e-mails that remain in the inbox, usually stored in the time-sequenced order of when it reached the recipient.

9.12 Blogs

Blogs are personal Internet diaries or journals, which have become popular in recent years. These are updated very frequently (often daily) and carry the author's observations and views on topics of her/his interest. They also include views and analysis of recent phenomena, favourite links, etc. Blogs typically have no single theme or structure. They are simple to develop and carry the personal style of the author. Some blogs are intended for family and friends, while others are meant to reach a much wider audience. The latter includes blogs that promote products, provide technical information or research results, or have political messages. Blogs written by eminent personalities in the field of art and science have large readerships.

To summarize, due to a wide range of applications and advantages, e-mail has gained wide popularity. It is now one of the most commonly used methods of communication. It is a very effective and useful tool in today's world, when time is a critical factor. E-mails have vastly diminished our dependence on the vagaries of the post office or courier service. However, it is not enough just to know how to write an e-mail. It is critical to know the rules of writing and managing e-mails effectively. If scientists and engineers around the globe become adept in the use of this new medium of communication, it would facilitate a more effective exchange of information and bring them closer.

10

Observing the Code of Gender Neutral Language

Language bears no bias
against any particular gender.
It is always impartial, always neutral.

L ANGUAGE is the basic tool of communication for everybody, irrespective of gender. In the process of communication, the words used should not reflect any prejudice or bias against any person based on gender. However, traditional language, as it is used even today, sometimes ignores or demeans the position of women. It is essential, therefore, to take care to use gender neutral language.

10.1 Awareness of the Social Effect of Language

The study of the psychological and social impact of gender neutral language began with the Sapir-Whorf hypothesis, named after two American linguists, Edward Sapir and Benjamin Lee Whorf. The hypothesis states that the thought processes of human beings are modified by the language they use. According to this hypothesis, a language is a particular mould in which thought categories are cast and where they get a particular shape and form. This is also referred to as linguistic determinism, which implies that our thinking is determined by our language. Marshall McLuhan, in his books,

The Gutenberg Galaxy (1962) and *Understanding Media* (1964), reinforces these views. McLuhan states that the media may bring about fundamental changes in society and also in the human psyche. 'Medium is the message', he states unambiguously.

Since then, the awareness of the social effect of language has grown. Human beings are social animals, living in a society in the midst of different social activities and getting influenced by the language they use.

However, there are those who do not agree with this perspective. They doubt if there is anything such as gender neutral language. According to them, this entire discussion may be politically motivated. They argue that even in cases where the language used is not gender neutral, instead of taking recourse to persuasion or enforcement, other means should be sought. Their position is that public attitude towards the issue should be changed gradually. They believe that the best recourse is to allow language to develop organically.

The Early Days: 'Man' as a Generic Term

Advocates of gender neutral language argue that traditional language fails to reflect the presence and position of women in society adequately. Today, the use of a simple word like 'man' can start a controversy. The word 'man' was once a generic term (applied to a whole kind or class) referring to all human beings irrespective of their gender.

The English Renaissance (1500–1650) saw the introduction of various Latin words (in classical and medieval forms) to the English language. The Industrial Age, spanning the 17th and 18th centuries, experienced the need to form new words. These were again borrowed from Latin or coined from Latin roots and affixes.

The grammar used at that time was meant to help boys in their study of Latin. During this period, very few women were given any education. The books written were for men and the grammar of that period focused on them. It was the product of a male-centric world. In the backdrop of a male-dominated society, the word 'man' became a generic term, referring to all human beings.

Narrowing of the Meaning of the Word 'Man' and its Impact

In course of time, the meaning of the word 'man' gradually narrowed. It began to refer to adult male human beings only. The grammar changed, too, reflecting this trend. Masculine pronouns, such as 'he', his', him', etc., did not refer to persons of both genders any longer.

This was when voices in favour of women's rights were being raised. The main demands were for equal treatment of women and men under the law and the right to franchise for women. A resolution towards this was passed after a long debate at the first Women's Rights Convention at the Seneca Halls, New York. At the session, 68 men and 32 women signed the 'Declaration of Sentiments', which included twelve resolutions based on the 'Declaration of Independence'. This process culminated in the 19th Amendment to the American Constitution, which granted women the right to franchise. The amendment was passed on the 26th of August, 1919.

These views were being reflected in society too. The realization dawned that there was a perceptible change in the roles of men and women. It was felt that since language and society were a reflection of each other, it was time to make appropriate modifications in language to reflect this change. As a result, gender neutral language began gaining the support of textbook publishers. Professional groups, such as the American Psychological Association and the Associated Press, also provided encouragement. Newspapers like *The New York Times* and *The Wall Street Journal* also discouraged the use of gender biased language.

Moving Away from the Generic Use of the Word 'Man'

This trend continues today and is gaining momentum. The language of today's multicultural world is discarding the limitations imposed by patriarchal society.

It is not difficult to write a text in gender neutral language. The challenge is to realize and accept the need to write in gender neutral language. Once that mental barrier is crossed, following a few simple guidelines is all that is needed.

Different methods can be used to make the language of a text gender neutral. These are described in the subsequent sections.

Using Substitute Words for Masculine Nouns

In gender neutral language, care should be taken to replace the word 'man' with prefixes, suffixes, or other combinations. Some examples are:

Avoid	*Use*
mankind	human race or humanity
manhood	adulthood or maturity
founding fathers	founders
common man	the average person
man made	manufactured, machine-made, synthetic

Replacing 'Man' as a Prefix

Earlier, almost always 'man' was used as a prefix (or otherwise) to signify the occupation. Nowadays, when the person holding the post could be a woman, the following substitutions can be made.

Avoid	*Use*
anchorman	anchor
businessman	business executive, business manager
chairman	chairperson, chair
clergyman	cleric, member of the clergy
common man	average person, ordinary person
craftsman	artisan
craftsmanship	artisanship
fireman	fire fighter
fisherman	angler
forefathers	ancestors
foreman	supervisor, superintendent

freshmen	first year students, new comers
gentleman's agreement	honourable (or informal) agreement
manpower	human resources, work force
master bedroom	largest bedroom
policeman	police officer
proprietor	owner
spokesman	spokesperson
steward	flight attendant
workmen	workers

Avoiding Feminine Suffixes

When using gender neutral language, feminine suffixes, such as -ess, -ette, and -trix, must be avoided. Nouns without these attachments bring the usage of the two genders at par. In the following examples, the nouns on the right are generic and can be used for both males and females:

Avoid	*Use*
poetess	poet
authoress	author
executrix	executor
usherette	usher
administratrix	administrator
actress	actor

Solving the Pronoun Problem

The generic use of the pronouns 'he', 'his', 'him', violates the principle of gender neutral language. These pronouns should be used to indicate males only. The following methods can be adopted to avoid the use of generic male pronouns.

Changing to the Plural Form

Here, the masculine noun is replaced by its plural form. For example:

Avoid: An engineer must pay attention to his work.
Use: Engineers must pay attention to their work.

Instead of: If a mechanic is negligent in his work, he will be punished.

Use: Mechanics negligent in their work will be punished.

Changing into Passive Voice

In this method, the sentence is changed to passive voice to avoid the use of the masculine pronoun. For example:

Avoid: He has returned the equipment today.

Use: The equipment has been returned today.

Instead of: The student could not solve the problem the teacher gave him.

Use: The problem given by the teacher could not be solved by the student.

However, it is best to use this form sparingly. Excessive use of the passive voice adversely affects the crispness of the text. Word-processing software such as Microsoft Word can be used to highlight instances where passive voice is used in a text.

Using Indefinite Pronouns and Articles

Often the generic use of masculine pronouns is avoided through the use of indefinite pronouns and articles. This is illustrated below:

Avoid: A man wanting his arrears must meet the accountant.

Use: Anyone wanting arrears must meet the accountant.

Eliminating Pronouns by Rewriting the Sentences

If none of the above methods help one to avoid the masculine pronoun, it is best to rewrite the sentence completely. For example:

Avoid: A good engineer performs the duty allotted to him efficiently.

Use: A good engineer performs the allotted duties efficiently.

Using 'she/he' as a Substitute for 'he' or 'she'

Some authors suggest the use of 's/he' as a substitute for 'he' or 'she'. However, 's/he' is not a valid word. Also, it is awkward to read. Therefore, wherever it is necessary, use she/he, him/her, he or she, etc.

Discarding Gender Bias when Referring to Pairs

A very subtle distinction may creep in while referring to persons of both genders in pairs. This usage puts males before females. However, in the context of gender neutral language, this usage may suggest that the females are just an add-on, or that they have been included as an after-thought. Varying the order of the words can solve this problem.

Some examples of such pairs are:

Avoid	*Use*
man and woman	woman and man
father and mother	mother and father
husband and wife	wife and husband
boy and girl	girl and boy

To refer to gentlemen and ladies, the expression 'ladies and gentlemen' is already in popular use.

Using Gender Neutral Language when Addressing unknown Persons

When writing to persons whose gender is unknown, it is more appropriate to write Dear Professor, Dear Doctor, or Dear Editor, etc. instead of Dear Sir/Madam.

Similarly, it is better to use a woman's first name instead of her husband's. The use of 'Ms' instead of 'Miss' or 'Mrs' is also recommended. This rule can be followed at all times, even when the marital status of the woman is known.

10.2 Gender Specific Usage: Driven by Readability, not Societal Norms

The use of the above guidelines will make the text truly gender neutral. However, it is necessary to ensure that this does not make the language clumsy. The overriding requirement for scientific texts is crispness and readability. A long-winded text adversely affects the clarity of the communication.

Hence, gender specific language is still used in many books and magazines. Its usage, however, is driven by a completely different intent and perspective. It has nothing to do with societal norms or the transformation that society has undergone. It is motivated by the sole purpose of keeping the language crisp and avoiding clumsiness in the text.

Hence, the decision to use gender neutral language in a scientific or technical text must be taken with care. Gender neutral language should be used only if it does not affect readability adversely. If adhering to a certain guideline significantly affects the crispness of the text, it should be avoided.

To conclude, the use of language that is not gender neutral comes, in most cases, very naturally and not because of any intended gender bias. It is used because of long-standing habit and tradition. Hence, what is needed is a little caution. If one realizes the necessity and importance of writing in gender neutral language, it can be done easily.

A language, just like human society, is never static. Language holds a mirror to society and reflects the changes it undergoes. In scientific and engineering texts, however, the overriding requirement is for crispness and readability and the use of gender neutral language must be balanced with it.

11

Look Before You Write: Avoid the Pitfalls

*If language is not correct
then what is said is not meant,
then what ought to be done
remains undone.*

—Ernest Gowers

A scientist or an engineer must write in flawless English and avoid the pitfalls of grammatical and spelling errors. This chapter deals with these pitfalls, but is not meant to be a treatise on English grammar or spelling. Its purpose is to make writers more cautious.

The list of such pitfalls, stated in the sections below, is not comprehensive, but is indicative of common errors that must be avoided. It will make writers more cautious about the use of appropriate words and phrases and will help them avoid the spelling and grammatical errors that are commonly part of scientific and engineering texts today.

These pitfalls are listed below.

Difference between 'disinterested' and 'uninterested'

'Disinterested' implies being impartial. For example:

Correct: A journal reviewer should be disinterested.

'Uninterested' means lack of interest, indifferent. For example:

Correct: He is uninterested in organic chemistry.

Difference between 'irreparable' and 'not repairable'

While 'irreparable' indicates that something is beyond rectification or ammendment, 'not repairable' signifies that something cannot be made ready to serve/use.

The use of these two words in the following sentences makes the difference in their meanings clear.

Correct: The loss caused by John's death is irreparable.

Correct: This broken chassis is not repairable.

Difference between 'after' and 'in'

The use of 'after' signifies a past space of time, while 'in' refers to a future space of time. This difference is highlighted in the sentences below:

Incorrect: The expert will come after a few days.

Correct: The expert will come in a few days.

When referring to the past, the correct and quite obvious expression is:

Correct: The expert returned after a few days.

Use of 'neither/nor'

If 'neither' or 'nor' comes at the start of a sentence, the verb is placed before the subject. For example:

Incorrect: Neither you will go to a new company, nor you will resign from this job.

Correct: Neither will you go to a new company, nor will you resign from this job.

Difference between 'of course' and 'certainly'

'Of course' can be used when what is being mentioned denotes a natural or inevitable consequence. 'Certainly' implies a forceful stance. The sentences below bring out this difference.

Incorrect: I shall of course go to the lab tomorrow.

Correct: I shall certainly go the lab tomorrow.

Correct: Of course, the flame was extinguished due to the lack of oxygen.

Using don't', won't, shouldn't, and etc.

Contractions, such as 'don't', 'won't', 'shouldn't', etc., should be avoided in formal writing. While 'etc.' (et cetera) is used correctly in the previous sentence, it is recommended that it not be used indiscriminately. This is because 'etc.' may sometimes make the meaning of a sentence vague.

For example, in the sentence below it is not clear which or how many additional locations the company's branches are located at.

Ambiguous: Our company has branches in France, Germany, etc.

Difference between 'mutual' and 'common'

The word 'mutual' means reciprocal (or, given and received). 'Common' means 'shared by two or more persons.' The two sentences below bring out the appropriate usage.

Incorrect: Jay is a mutual friend of Albert and Roy.

Correct: Jay is a common friend of Albert and Roy.

Correct: Unless there is a mutual understanding between the leaders of the two companies, this alliance will not last.

Appropriate Usage of Phrases

Phrases and idioms, when used in a sentence, should be written without any change. For example:

Incorrect: Although they worked with heart and soul, they did not succeed.

Correct: Although they worked heart and soul, they could not succeed.

Incorrect: I must complete the experiment by hook or crook.

Correct: I must complete the experiment by hook or by crook.

Appropriate Use of 'its', 'hers', 'yours', 'ours', and 'theirs'

The words 'its', 'hers, 'yours', 'ours', and 'theirs' are possessive. However, these should never be written with the apostrophe sign (').

Incorrect: It's boiling point is 100° Celsius.
Correct: Its boiling point is 100° Celsius.

Appropriate Position of Adjectives

If qualifying phrases are enlarged by an adjective, the adjective is placed after the noun it qualifies. For example:

Incorrect: To reach the site one has to cross a twenty feet wide road.
Correct: To reach the site, one has to cross a road twenty feet wide.

Appropriate Order of Pronouns

When there are pronouns indicating first person, second person, and third person, the second person should come first, third person next, and the first person last. However, while making an admission of guilt, the first person should come first.

Incorrect: Me, you, and Amy have been placed on the same team.
Correct: You, Amy, and I have been placed on the same team.

Difference between 'say', 'tell', and 'speak'

There is a difference in the usage of the words 'say', 'tell', and 'speak'. 'Say' is used to assert or declare. It is used when referring to the actual words uttered by someone.

'Tell' is used to inform, narrate, or command. It is commonly used in indirect speech when the sentence contains an indirect object.

'Speak' refers to uttered words.

The difference between 'said' and 'told' is brought out in the following two examples:

Incorrect: The scientist told, 'I shall return'.
Correct: The scientist said, 'I shall return'.

Incorrect: The scientist said to me that he would return.
Correct: The scientist told me that he would return.

Use of 'many a'

'Many a' should always be followed by a singular noun and singular verb. However, if 'many a' is followed by two nouns with different meanings, the verb should definitely be plural. For example:

Correct: Many a man was working at the site.
Correct: Many a man and woman were working at the site.

Use of Nouns Preceded by 'each', 'every', or 'no'

Singular nouns connected by 'and' that are preceded by 'each', 'every', or 'no', are followed by verbs in the singular number. For example:

Incorrect: Every metal and every alloy react with this unique
              substance.
Correct: Every metal and every alloy reacts with this unique
              substance.

Difference between 'until' and 'up to'

'Until' is always used to indicate time, while 'up to' is used to denote place. For example:

Incorrect: The author wrote the paper up to 10 o'clock.
Correct: The author wrote the paper until 10 o'clock.

Difference between 'a number of' and 'the number of'

The expression 'a number of' always takes a plural verb. In contrast, 'the number of' always takes a singular verb. For example:

Incorrect: A number of engineers was present on that day.
Correct: A number of engineers were present on that day.

Incorrect: The number of engineers present were very small.
Correct: The number of engineers present was very small.

Use of Correlative Words

Each member of a pair of correlative words should be placed before the same parts of speech. This is illustrated by the example below:

Incorrect: Yield loss is not only hard to remove, but also to detect.
Correct: Yield loss is hard not only to remove, but also to detect.

Use of 'no sooner', 'scarcely, and 'hardly'

In clauses beginning with 'no sooner', 'scarcely', 'hardly', etc., the auxiliary verb precedes the subject. For example:

Incorrect: Hardly they had left when the delegation arrived.
Correct: Hardly had they left when the delegation arrived.

Use of 'the more...the more'

When there are two clauses joined by a pair of correlative conjunctions such as 'the more...the more', the subject in the second clause is placed after the verb or the auxiliary verb. For example:

Incorrect: The more you simulate, the more you will make the design robust.
Correct: The more you simulate, the more will you make the design robust.

Use of 'who', 'whom', 'whose', 'which', and 'that'

The relative pronouns 'who, 'whom', 'whose', 'which', and 'that' should be placed just after the antecedent. For example:

Incorrect: I have read Bose's papers who was a great scientist.
Correct: I have read the papers of Bose, who was a great scientist.

Use of Single or Plural Verbs in Collective Words

There could be confusion in one's mind regarding the use of singular or plural verbs when using collective words such as committee, team, etc.

While there is no rigid rule, the common practice is to use the plural verb when the emphasis is on individual members. However, when the body as a whole is considered, the singular verb is to be used. For example:

Correct: The technical committee was unable to agree to the proposal.

Incomplete Comparisons

Incomplete comparisons create ambiguity and may convey a meaning other than the one intended. Hence, these should be avoided. For example:

Ambiguous sentence: She ranks Sue higher than Susan.

This sentence is ambiguous because it may mean:
She ranks Sue higher than she ranks Susan.

It could also mean:
Susan ranks Sue high, but she ranks Sue higher than Susan ranks Sue.

Use of 'in order that'

'In order that' should be followed by 'may' or 'might'. For example:

Correct: In order that no chemical reaction may arise…

Use of 'therefore' and 'only'

'Therefore' is used after a word or phrase that is emphasized. This is the sole function of 'therefore'. The word 'only' should be placed immediately before the word it is intended to modify. For example:

Correct: Silicon dioxide showed all the required characteristics. Therefore, it can be used as a dielectric.

Correct: This, however, does not imply that only Silicon dioxide can serve as the dielectric.

Referring to Anything More than One

Anything more than one must be taken as a whole when it has to be written in plural form. For example:

Correct: The biologist poured one-and-a-half ounces of the substance into the jar.

Correct: The author has completed three-fourths of the book.

Here 'three-fourths' mean three parts out of four. That is why the plural form has been used.

Referring to Physical and Mental Feelings

When referring to the physical and mental feelings of a person, the appropriate articles should be used. Thus, indefinite articles like 'a' or 'an' are used before 'rest', 'cold', 'headache', 'temper', 'rage', etc. For example:

Correct: We shall take a rest here for 30 minutes.

Correct: The manager was in a temper over this outrageous error.

Correct: I have a headache after working on this paper for so long.

Use of Verbs such as 'appoint', 'consider', 'think', 'call'

Verbs such as 'appoint', 'consider', 'think', 'call', etc., are not followed by 'as'. For example:

Incorrect: He was appointed as President of the Chemical Society.

Correct: He was appointed President of the Chemical Society.

Incorrect: I consider him as the fittest man for the job.

Correct: I consider him the fittest man for the job.

Use of the Verbs 'regard', 'depict', 'define'

Verbs such as 'regard', 'depict', 'define' and 'mention' are followed by 'as'. For example:

Correct: I regard Joseph as my mentor and role model in the organization.

Correct: In the paper, this variable has been defined as a string.

Use of Apostrophe ('s)

The apostrophe ('s) is added only in the case of animate objects. It is not used in the case of inanimate objects except after expressions that denote time or distance. For example:

Correct: He will be back in an hour's time.

Correct: His hostel is a stone's throw from the Institute.

Apostrophe is also used in phrases such as:

- a bird's eye view
- at his wit's end
- to his heart's content

The Importance of Punctuation Marks

The absence of the correct punctuation mark, such as a comma, can create havoc with the meaning of sentence. Once a very innocuous (not intended to offend) sentence in a newspaper led to a furore among several organizations. The sentence read: 'Woman without her man is helpless'.

The editor, who was well versed in punctuation and also human psychology, came out with an erratum the next day, saying that the sentence was the result of the printer's oversight. He pacified the irate organizations by apologizing and saying that the original version was: 'Woman—without her, man is helpless.'

Use of Present Participle

A participle is also called a verbal adjective, that is, a verb form doing the work of an adjective. A present participle denotes unfinished work and ends in '-ing'. A present participle should not be used with a verb denoting a past action. For example:

Incorrect: He started the simulation on Friday and finishing on Sunday.

Correct: He started the simulation on Friday and finished on Sunday.

Difference between 'partly' and 'partially'

Partly and partially are not synonymous terms and should not be used interchangeably. The dictionary meaning of 'partly' is 'to some extent' (that is, not fully). 'Partially' means just forming a part that is not complete. However, the meanings seem to overlap. Partly is the adverb of 'part', and 'partially' of partial. For example:

'He is partly responsible for the experiment's failure'.

This means, he is not fully responsible for the experiment failure; somebody else is also responsible.

However, in the case of 'partial', as in the partial eclipse of the sun, it implies that the rest of the sun is clear. Hence, partly suggests the presence of something or somebody else. Partial only means 'not complete', without any reference to anything else.

Use of 'pair', 'dozen', 'score', 'hundred', 'thousand', etc.

After numerals, 'pair', 'dozen', 'score', 'hundred', 'thousand', etc., take a singular form. For example:

Incorrect: The process needs to be repeated for all three pairs of setups.

Correct: The process needs to be repeated for all three pair of setups.

However, when numeral adjectives that are not defined or unspecified come before such words, the plural form is used. For example:

Correct: The process needs to be repeated for all pairs of setups.

Correct: The process needs to be repeated for hundreds of pairs of setups.

Difference between 'little', 'small' and 'few'

'Little' denotes quantity, 'small' denotes size, while 'few' denotes a number. For example:

Correct: There is little acid left in the jar.

Correct: The small round object attracted the attention of the scientist.

Correct: There are only a few jars in the lab.

Use of the Word 'Machinery'

The term 'machinery' (not machineries) should be used when referring to a composite set of machines. For example:

Incorrect: India imports heavy machineries.
Correct: India imports heavy machinery.

Difference between 'whether' and 'whether or not'

'Whether' should be used to introduce at least two alternatives, either stated or implied. In contrast 'whether or not' means 'regardless of whether'. For example:

Incorrect: I am not sure whether or not to repeat the experiment.

Some correct usages of 'whether' are stated below:

Correct: I am not sure whether to repeat the experiment.
Correct: I am not sure whether I should repeat the experiment or use a different one.

Some correct usages of 'whether or not' are provided below:

Correct: Whether or not I repeat the experiment, I will leave the laboratory late tonight.
Correct: Whether or not the results are positive, I will repeat the experiment.

Use of the Expression 'the same'

The expression 'the same' should not be used as a substitute for a personal pronoun. For example:

Incorrect: When you have read the document, please return the same to me.
Correct: When you have read the document, please return it to me.

Use of 'Comprise'

'To comprise' implies 'to contain' or 'to consist of'. It is not a synonym for 'to compose'. For example:

Incorrect: The physics book is comprised of ten chapters.
Correct: The physics book comprises ten chapters.

Incorrect: Our research was comprised of three stages.
Correct: Our research comprised three stages.

Use of 'Articles'

The articles 'a' and 'an' should be chosen according to the pronunciation of the words or abbreviations they precede. For example:

Correct: A nuclear magnetic resonance spectrometer
Correct: An NMR spectrometer

The proper article must be chosen to precede B.A., B.S., M.A., M.S., and Ph.D., according to pronunciation of the first letter. For example:

Correct: He has a B.S. degree.
Correct: He has an M.S. degree.
Correct: He has a Ph.D.

The definite article 'the' is not used before the following nouns when they are used for their usual or primary purpose: school, college, university, church, prison, hospital.

For example, when an engineer is going to her/his own office, she/he will say: 'I am going to office'. But when a visitor goes to her/his office, the visitor will say: 'I am going to the office'.

Gerund versus Infinitive

Very often there is confusion regarding whether to use a gerund or an infinitive in a particular sentence. A gerund is also called a verbal noun because it does the work of a noun too. It is formed by adding '-ing' to a verb. For example: completing, analyzing, etc. The rules are simple enough: Verbs that come immediately after prepositions must be in gerund form. For example:

Incorrect: Are you interested to get your results published?
Correct: Are you interested in getting your results published?

An infinitive consists of two words: 'to' and a verb. For example: to complete, to analyse, etc.

After the following phrases, however, a gerund and not an infinitive should be used, such as: 'with a view to', 'look forward to', etc. For example:

Incorrect: He went to Kolkata with a view to assess his customer's needs.

Correct: He went to Kolkata with a view to assessing his customer's needs.

Incorrect: I am looking forward to meet you.

Correct: I am looking forward to meeting you.

Use of Perplexing Prepositions

There are rules for the proper use of prepositions. However, the perplexing prepositions do not adhere to set rules. They follow usage and conventions. This is why, at times, it becomes difficult for a person not conversant with these usages and conventions to pick up the right preposition.

Here are a few examples of some perplexing prepositions. Rules have been cited wherever possible without entering into the intricacies of grammar. Broadly speaking, a preposition points out the relation of one word with another.

Here are some examples of perplexing prepositions:

(1A) This is the subject of discussion.

Here, the use of the preposition 'of' implies that the discussion is going on. Let us compare this with the following sentence:

(1B) This is the subject for discussion.

Here, the preposition 'for' implies that the discussion has not yet begun.

When used with verbs that denote verbal communication, such as 'talking', 'speaking', etc., the prepositions 'to' and 'for' convey different meanings. The following two examples illustrate the difference.

(2A) We spoke to the manager.

Here, the preposition 'to' conveys the meaning that the subject (we) did all the speaking. Let us contrast the above with the following sentence:

(2B) We spoke with the manager.

The preposition 'with' conveys the meaning that there was a two-way discussion between the two parties: the subject (we) and the manager.

Here are some more examples of perplexing prepositions:

(3A) They rejoiced on their own success.
(3B) They rejoiced at the success of their friend.

(4A) Jim called at my office yesterday.
(4B) Jim called on me yesterday.

(5A) The student is playing at the computer.
(5B) The student is playing on an instrument.

(6A) I was gazing at the stars to understand their relative positions.
(6B) I was gazing upon the beautiful combination of colours.

(7A) May God protect you from harm.
(7B) I shall protect you against any adversity.

(8A) The supervisor is annoyed by the rude behaviour.
(8B) The supervisor is annoyed with your colleague.

(9A) This is beyond my understanding.
(9B) You look beyond your age.

(10A) My friend put me in danger.
(10B) My friend put me to shame.

(11A) The bank lends money at a low rate of interest.
(11B) The banks lend money on good security.

(12A) The man stared at me.
(12B) The man stared me in the face.

(13A) The manager was angry with the employee.
(13B) The manager was angry at the results.

Words like 'cause', 'use', 'need', 'reason', etc. normally take the preposition 'of' after them. However, if they are preceded by 'any' or 'no', they take the preposition 'for'. The following pairs illustrate this:

(14A) Is there any cause for delay?

(14B) This is the cause of my delay.

(15A) There is no reason for your failure.

(15B) This was the reason of your failure.

(16A) I have no need for his help.

(16B) What is the need of your going there?

Finally, here are a few more examples of prepositions, which may appear perplexing, but are correct.

- Do not have medicine on an empty stomach.
- Mr Kumar is not on the review committee.
- The book has passed through ten editions.
- I can see through the whole plot.
- The project is running to time.
- He borrowed money on the deposits of his colleague.
- Do not rejoice over the misfortune of others.
- We closed with (accepted) the offer.
- The manager wanted to discuss the problem with the members of the Board of Directors. (Note: Discuss 'about' the problem should not be used).
- Sometimes it becomes difficult to cope with the problems. (Note: 'Cope up with' should not be used).

Commonly Misspelled Words

Care should be taken to ensure that there are no spelling errors in the text. Incorrect spelling can mar the authenticity of any document. This has been dealt with here, along with grammatical errors, because grammar and spelling are an integral part of any scientific or technical writing and, hence, deserve due importance. It is possible to avoid spelling errors with a little care and patience.

The list that follows is not intended to be exhaustive. It has been provided to make writers cautious of the pitfalls they may face with spelling. The words are not necessarily scientific terms—but are the ones that are often used by scientists and engineers in their communication. These are, thus, a sample of words that are used often, yet are confusing to the author and prone to errors.

Abbreviate
Abhorrent
Accede
Accessible
Accessory
Accommodate
Accomplice
Accomplish
Accredit
Achieve
Acknowledge
Acquitted
Addendum
Address
Adjudge
Adverse
Aggregate
Agreeable
Alignment
All Right
Allotted
Allied
Allocate
Allotment
Allowance
Alter
Alternate
Amateur
Ambiguous

Anaesthetic
Annexation
Antiseptic
Aperture
Apparatus
Appeal
Argument
Artillery
Ascertain
Ascetic
Assessment
Asynchronous
Attenuation
Author
Avionics

Bakelite
Bankruptcy
Baulk
Beginning
Believe
Benefit
Beneficial
Besiege
Bibliophile
Bicarbonate
Bizarre
Blitzkrieg
Boundary
Breathe

Brevity
Brief
Budget
Bulletin

Cadence
Calorie
Camouflage
Capillary
Capacious
Career
Carrier
Cassette
Catalogue
Ceiling
Chargeable
Chassis
Circular
Circumference
Claimant
Cliche
Coaxial
Coincide
Collapse
Colloquial
Collusion
Colour
Column
Committee
Communicate
Compass
Compressor
Concurrence
Conscious
Consensus
Convalescence

Convenience
Corollary
Cosmonaut
Coulomb
Critique
Curious

Dental
Deceive
Decibel
Decision
Deficient
Demurrage
Descendant
Diameter
Diaphragm
Dictionary
Dimension
Director
Disease
Disguise
Dispatch
Dissolve
Draught
Drought

Eclipse
Efficient
Embarrass
Emission
Encumbrance
Endurability
Equilibrium
Erroneous
Et cetera
Exacerbate
Exaggerate

Exempted

Exhaust

Exhilarate

Existence

Extension

Fahrenheit

Farther

Fascinate

Feasible

Feminine

Fluorescent

Forfeit

Forty

Freeze

Fruition

Gadget

Gaseous

Gauge

Gauss

Gazette

Genealogy

Grammar

Grateful

Grievance

Guarantee

Guide

Gynaecology

Haematology

Haemorrhage

Hand-over

Hereditary

Harmonize

Hawkeyed

Hygiene

Inaccessible

Incandescent

Indefatigable

Independent

Indigenous

Infinitesimal

Miscellaneous

Misconstrue

Moderator

Momentous

Movable

Muscle

Multitudinous

Myriad

Necessary

Negligible

Neuter

Ninth

Noticeable

Nugatory

Obligatory

Observance

Obsolescence

Obsequious

Obsolete

Occasion

Occurred

Occurrence

Omitted

Opportunity

Oscillate

Oscilloscope

Pandemic

Panegyric

Paradigm
Parallelogram
Parlous
Patience
Perceive
Permanent
Permeability
Personnel
Persuade
Physique
Pillar
Pique
Plausible
Pneumonia
Possession
Precede
Predecessor
Preliminary
Prevail
Privilege
Programme
Pronounce
Pronunciation
Proprietor
Psyche
Punctual
Puncture
Pursue
Pursuit

Quadrant
Querulous
Quotation
Quotient

Receipt
Receive

Recession
Referee
Relieve
Remission
Repentance
Repetition
Replaceable
Resistance
Rheostat
Rigorous

Saturation
Schedule
Schematic
Scholastic
Sciatica
Scrupulous
Secondment
Secede
Secretary
Sedentary
Segregate
Seize
Separate
Sinusoidal
Siege
Sieve
Sizeable
Skeleton
Susceptible
Synchronous

Tariff
Telecommunication
Territory
Tertiary
Thorough

Threshold	Variety
Tolerance	Vector
Toxicity	Ventricle
Transferred	Veracity
Transmitter	Verbiage
Trespass	Vestige
Trigonometry	Vexed
Truly	Vis-a-vis
Tuition	Viscous
Tutelage	Volte-face
Twelfth	
	Wattage
Unique	Wield
Unparalleled	Wedge
	Withered
Vaccinate	
Vacillation	Yield
Vacuum	

To conclude, this chapter helps the scientist to write in correct English. While most scientists and engineers focus on the technical content (and rightly so), it is imperative that the grammar and the spelling in the text are accurate. This is a fundamental requirement of any well-written scientific or engineering text.

Appendix A aids this process further by illustrating the role that 'roots' play in building and using the right words. Appendix B expands the focus on avoidance of pitfalls, by providing several examples of words and expressions that appear similar, but have very different meanings.

Conclusion

A document or a speech
should be aesthetic and inviting;
not a jungle of jumbled words and sounds.

COMMUNICATION is the foundation of the inter-connected world. It is the single most important factor that has caused the decline of distance. Nowadays, life in the global village has become more fast-paced than ever before. It has become essential for people across the world to be able to indulge in effective and instant—written as well as verbal—communication. This is as critical between individuals in the same location as it is between those across varying cultures. In this day and age, the recipients of any type of communication do not usually have an abundance of time. Hence, they want to have the right facts presented succinctly and in the expected order. In addition, scores of individuals in various organizations have realized the increased importance of teaming and interaction in this age of globalization.

The need is for everyone to be in lock step, driven by a common goal. This requires a multi-way communication channel that keeps all those concerned synchronized on vital matters and trends. Effective communication techniques—verbal and written—go a long way in making this happen. What are required are clarity of thinking and expression, and an understanding of the various genres of communication techniques. The specific rules and tools concerning communication vary depending upon whether it is a technical report, an e-mail, a technical paper, a précis of a scientific matter, or a speech.

An appreciation of the uniqueness of each method can greatly help increase the effectiveness and instantaneity of communication in any organization. It can thereby assist scientists and engineers in advancing their professional careers. *Clarity* is the fundamental source of achieving this effectiveness in communication. Simultaneously, it is *technology* that brings in the instantaneity of communication.

Clarity is built brick-by-brick; it starts with the words used. The choice of the right words should be a conscious act rather than a random one. There are eight guiding principles stated in this book that aid this selection. These range from the usage of simple words without redundancy, to selecting precise words that conform to the intended meaning in the technical world. The vocabulary of science and engineering is growing and changing everyday. Hence, using words that are driven by precision (not by style or pompousness) is important. This is possible only if the author is fully aware of the exact meanings of the words used. For example, the word 'discharge' in traditional English means 'release' or 'dismiss'. However, in technical communication, 'discharge' refers to the loss of stored charge.

A similar situation arises while using synonyms that have apparently similar meanings. In non-scientific texts, synonyms can be used interchangeably; this is not the case with technical communication. The specific nuances in the meanings, which differentiate one word from another, should be always borne in mind. For example, words, such as 'cause', 'reason', and 'purpose', appear to have the same meaning on the surface. However, they have very distinct meanings in reality. If one delves into the precise meaning of each, the differences come out easily. The word 'cause' refers to something that makes a thing happen. For example: 'What was the cause of the failure?' The word 'reason' furnishes explanation. For example: 'He gave reasons why the sales dipped.' Finally, 'purpose' states why something is being done. For example: 'The purpose of the simulation is to test the property of the chemical under various conditions.'

When writing texts of a technical nature, words should be carefully threaded together to form short sentences. Ten guidelines, which aid this process, have been provided in this book. Once again, simplicity and brevity are fundamental for increased readability. Sentences must be short, positive, and direct. They should not attempt to convey multiple messages in a convoluted manner. A good knowledge of grammar, especially voice and tense, comes in very handy in concatenating the words.

Sentences form the next layer of bricks; they build paragraphs in a step-by-step manner. Too often the succinctness of paragraphs is not respected. Long paragraphs often tend to cover multiple themes leading to confusion in the minds of the readers; they can also go on in a rambling manner, which makes the readers lose complete interest. It is critical to ensure that a paragraph serves as a single unit. Sequencing paragraphs is also critical to enable a seamless transition from one subject to the next. After developing the paragraphs, a quick check using the fog index serves as a useful self-test to assess the degree of clarity.

With these fundamental building blocks in place, a scientist or engineer begins to develop the technical content. This could be a scientific paper, a technical report, or even an e-mail. In any scientific or engineering text, the author has an assortment of data. The critical aspect of a technical study is to analyse the data and to present the results of the analysis. This requires organizing the body of knowledge into information and structuring these into separate sections. Adherence to an appropriate structure, and to the guidelines within each structural component, is of vital importance in this process.

A planned structure greatly improves the effectiveness of communicating the content to the readers. It lends a sharp edge to a document when it is complete and makes it attractive. Most importantly, a structured format is of great help to the readers. It helps them grasp the flow of ideas easily. Cues, pointers, headers, etc. act as pathfinders and indicate the structural relations amongst different aspects of the text. These, coupled with the right font sizes and font types, enhance the aesthetics of a technical document. This ensures that a document or slide looks like an attractive picture, drawing instant attention—not like a frightening jungle of jumbled words.

However, in technical communication, a planned structure is not sufficient. For such communication, equations, chemical reactions, symbols, standard deviations, etc. are usually necessary. These must follow international standards and guidelines. In addition, guidelines used across the world, for system modelling and simulation based on experimental data, should be adhered to. It is only after this has been accomplished that a reader, located thousands of miles away, will be able to appreciate the intent of the author.

Accuracy is of high importance in technical communication. This includes data sanctity, correct spelling, and proper grammar. Gaining expertise in scientific writing is, therefore, an involved process. It can be acquired in two ways: reading good scientific journals and meticulously observing the methods adopted by reputed scientists in their documents. A thorough acquaintance with state-of-the-art research methodologies and principles of scientific research can also be very useful.

Over the years, one form of communication has blended remarkably between the old and the new—speech. Hence, this book delves into hundreds and thousands of years of history to bring back some glimpses of effective speech communication. Incisive, carefully crafted, and motivational speeches by Brutus and Antony (from the works of Shakespeare), by Martin Luther King Jr., by Winston Churchill, and by Swami Vivekananda have stood the test of time. The essence of speech communication is that it must tell a story. It stands on the fundamental premise that it is dynamic. The speaker must be able to adapt the story continuously while gauging the mood of the audience, and must keep in mind that the benefits to the audience need to be communicated. The speech should not be stated from the viewpoint of the speaker. A well-conceived speech is usually threaded together by a storyline. Presentations must be interspersed with the diverse technological aids that are available today. A powerful speech, blended judiciously with graphics, audio, and video aids, can have a lasting effect on the audience.

All the aspects mentioned above, that lead to clarity of texts, have been dealt with in detail in different chapters of this book. The key point is that acquiring proficiency in high clarity communication is a critical skill. It is imperative for scientists and engineers to be well-versed in this skill to achieve success in their respective professional career.

Technology is a very important ally when meeting the increased need for efficient and instant communication. Graphical aids, audio and video equipments, automated tools, etc. have dual purposes. Used effectively, these serve to convey a message unambiguously. They also give the readers or the audience a welcome break from the monotony of a long text, and regenerate their attention. Use of technological aids is, however, not without its pitfalls. Judicious use of the appropriate tools is critical to ensure that the right message is communicated. However, indiscriminate use of a particular form may convey unintended results. For example, a pie chart and a line chart are not interchangeable. Neither are audio clips and video clips. Each tool and each form has its own purpose and place.

Where will communication evolve from here? Technology today is already blurring the lines between private and public spaces. What used to be personal views and opinions, written in emotion-packed diaries, are now making headlines as Internet diaries or blogs. This has thrown open the opportunity for people to express their viewpoints overtly—and for readers around the globe to savour these. Internet chat-rooms have replaced the close-knit community as people can debate about their views and find friends online now.

In this new world order, therefore, scientists and engineers must continuously adapt themselves to these trends of new technology. At the same time, they should not lose sight of the fundamental tenets of written and speech communication stated in this book. This combina-tion will give the readers the winning edge in their own professions.

Appendix A

The Role of Roots in Word Building

A good knowledge of the structure of words and how they are formed is of great help in strengthening one's vocabulary. Most words in the English language today have been borrowed from other languages. A majority of these words are of Latin and Greek origin. It is possible to understand the meaning of a word by studying its root along with its prefixes and suffixes.

The 'root' is the base element of a word. It contains the basic meaning of the word. Since prefixes and suffixes are attached to a root, they are referred to as affixes. A prefix is a word element that is placed in front of a root. A suffix is a word element that is placed after a root. A root, with the help of a prefix or suffix, changes the meaning of words. Like the root, the prefix and suffix also carry particular meanings. Hence, one who is conversant with the meaning of a root can easily guess the meaning of a word.

Here is a selective list of some words, with their roots and affixes.

Root/Affix	Meaning	Examples
a, ad	in addition to, towards, near	adverb, accompany, affix, alive
a, an	not, without	apolitical, apathy, anarchy
able	capable of being	manageable, unable
am, ami	love, like	amicable, amiable
ambul	to walk	amble, ambulance, somnambulist
anim	life, spirit, anger	animate, animosity
ann, annu	yearly	anniversary, annual, annuity
aero	air	aeroplane, aeronaut, aerobic

(Contd.)

(Contd.)

Root/Affix	Meaning	Examples
age	state of, rank	usage, marriage
amphi	round, on both sides	amphitheatre, amphibious
ance, ancy	state of being	elegance, assistance, discrepancy
ante	before	antechamber, antedate, antecedent
anti	against, opposite,	antisocial, antibody, antiseptic
aud, audi, aur	to hear	auditorium, audience, audiovisual
arch	ruler, chief	archbishop, archenemy, archduke
astro, astrom	star	astronomy, astrophysics, astrology
auto	self	autonomous, automatic, autobiography
bene	good, gentle	beneficial, benevolent, benefit
bi	two	biped, bicycle, bifurcate
bio	life	bioscience, biology, biography
bibli, biblio	book	bibliography, bibliophile
brev	short	brief, brevity, abbreviate
cede, ceed	go, move	recede, proceed, exceed
cent	hundred,	century, centimeter
chron	time	chronometer, chronology, synchronize
cardio, cardi	heart	cardiology, cardiograph
counter, contra	opposite, against	counteract, contradict, contrary
de	doing the opposite	detach, decrease, degenerate
demo	people, town	democracy, demagogue, demography
dic, dict, dit	say, speak	dictate, verdict, malediction
dis, dif	opposite of, separate, away	differ, divide, distrust, disconnect
dom	quality, realm, office	kingdom, thralldom, freedom
ence	state, quality, fact	competence, affluence
en	made of, make	wooden, woolen, frozen
equi	equal	equidistant, equilibrium, equitable
extra, exter	outside of, beyond	external, exterior, extracurricular
er, or	person who does something	driver, porter, sweeper
fy	make, do	glorify, magnify, exemplify
ful	having, containing	hopeful, helpful, plentiful
geo	earth	geology, geography
graph, gram	to draw, to write	graphic, lithograph, autograph
gor	to bring together	category, categorize
hetero	different, other	heterogeneous, heterodox, heterosexual
hyper	over, above	hypersensitive, hyperbolic, hyperactive

(Contd.)

(Contd.)

Root/Affix	Meaning	Examples
ic	akin to, alike	metallic, charismatic, emblematic
in, im, ir	undecided, stagnant	inaction, irresolute, impossible
infra	beneath, lower	infra-dig, infrared, infrastructure
inter	between, among	intermission, international, intermediate
intra	within	intramuscular, intravenous, intra-mural
intro	into, within	introspection, introvert
itis	illness	arthritis, appendicitis, hepatitia
ive	belonging to	prospective, automotive
ize	to become like	computerize, stabilize, customize
ject	to throw	eject, project, inject
low	inferior, restrained	low-grade, lowly, low-budget
macro	large, big	macroeconomics, macrocosm
mega	great	megastar, megalomania
ment	means, product, act	amazement, containment
meta	transformation	metamorphosis, metaphysics, metaphor
meter	measure	barometer, thermometer, speedometer
micro	small	microscope, microprocessor, microfilm
mid	middle	midnight, midday, midsummer
milli	thousand	millimetre, millipede, millisecond
mis	bad, wrong	misconduct, mispronounce, misinterpret
mini	small, short, least	minimum, minibus, miniskirt
mono	one	monopoly, monogamy, monologue
multi	many	multipurpose, multidimensional
naut	sailor	astronaut, cosmonaut, aeronaut
ness	state of	happiness, childishness
neuro	nerve	neurology, neurosurgery, neuron
non	not	noncommittal, nonentity, nonconformist
nym	name	pseudonym, acronym
ocracy	form of government	democracy, autocracy bureaucracy
octo	eight	octogenarian, octave, octopus
omni	all	omnipresent, omnivorous, omnipotent
ory	place for	laboratory, conservatory
osis	disease, illness	tuberculosis, cirrhosis, thrombosis
out	more, to do better	outwit, out-manoeuvre, outsmart
ous	characterized by, full of	monotonous, gluttonous, glorious
over	above	overeat, overcome, overwork

(Contd.)

(Contd.)

Root/Affix	Meaning	Examples
para	beside	paramilitary, paramedical, paratyphoid
ped, pod	foot	pedestrian, pedestal, podium
peri	round	perimeter, periscope, period
phobia	fear	arachnophobia, claustrophobia
poly	many	polyandry, polygamy, polyglot
post	after	postpone, postscript, postdated
port	carry	teleport, porter
pre	before	precede, premature
pro	before	prologue, propel, procrastination
pseudo	false	pseudonym, pseudo-intellectual
psych, psycho	mind	psychotherapy, psychoanalysis
retro	backwards	retrospect, retrogression, retroactive
semi	half	semiconductor, semifinal, semicircle
sub	under	sub-inspector, submerge, subordinate
super	over	supercede, superlative, superstructure
syn, sym	at the same time	synonym, synthesis, synchronous
techn	craft, skill	technology, technocrat, technician
tele	from a distance	telephone, telepathy, telegraph
tort	twist, change	contort, distort
trans	beyond, across	transcend, transform, transmit
tri	three	tricycle, triangle, tripod
theo	God	theology, theism, atheist
ty	condition of	enmity, exclusivity, longevity
uni	one	unity, union, unicycle
vice	in place of	vice-president, vice-chancellor
with	against	withstand, withdraw, withhold

Appendix B

Confusing Words and Expressions

The English language has various words and expressions that can confuse writers because they are similar in appearance. At such times, writers have to be careful about what they write. The confusion might arise because of various reasons. Sometimes writers have to make a selection amongst similar-sounding words called homonyms. Examples of such word-pairs are 'troop and troupe', 'ordinance and ordnance', etc. At other times, one may stumble upon words that appear similar, but are used differently. Examples in this category are words like 'sometime, sometimes, or some time', 'lie, lay, lain, or laid', etc. Proper care has to be taken when writers encounter such words that might appear similar in sound, spelling, or even meaning.

The list given below is not a comprehensive one. It is only intended to make writers more cautious when they come across similar words. The set of confusing words is given first, followed by examples that bring out the meaning of each of the words. The meaning of each of the words has been given.

Abdicate/Abrogate/Arrogate

- *Abdicate = relinquish, give up*
  The company *abdicated* all responsibility after the event.
- *Abrogate = formally annul*
  The chairperson refused to *abrogate* any inherent powers.
- *Arrogate = assume*
  The Board of Directors *arrogated* the responsibilities of the leadership team.

Adjacent/Adjoining

- *Adjacent = quite near but not attached to the main*
  The meeting took place in the *adjacent* building.
- *Adjoining = attached*
  The *adjoining* chart illustrates the example clearly.

Careful/Cautious

- *Careful = alert*
  The foreman warned the mechanic to be *careful* when handling machinery.
- *Cautious = guarded*
  The chairperson spoke with *cautious* optimism while describing the growth forecast.

Climatic/Climactic

- *Climatic = atmosphere in a particular place or time*
  The *climatic* condition of the region is quite suitable for travelling.
- *Climactic = reaching a climax*
  The *climactic* finale of the show was highly appreciated.

Complement/Compliment

- *Complement = parts of a whole unit*
  A good scientist and a good working environment *complement* each other.
- *Compliment = praise*
  The director was *complimented* for the excellent keynote address.

Contemptuous/Contemptible

- *Contemptuous = expressing contempt, showing strong dislike*
  The manager was *contemptuous* of those who shirked responsibilities.
- *Contemptible = deserving to be treated with contempt*
  The manager's distasteful behaviour in front of the whole team was really *contemptible*.

Cord/Chord

- *Cord = a tangible connection*
  The doctor took an X-ray test of my spinal *cord*.
- *Chord = emotional connection, musical notes sung together*
  The advertisement's appeal for help touched the right *chord* in my heart.

Credible/Creditable/Credulous

- *Credible = believable*
  What the employee communicated in the technical report was not *credible*.
- *Creditable = praiseworthy*
  The performance of the young engineer was quite *creditable*.
- *Credulous = gullible*
  People who are *credulous* by nature are often duped.

Definite/Definitive

- *Definite = certain, fixed*
  The candidate failed to give any *definite* answer to the question.
- *Definitive = conclusive, authoritative*
  There is no *definitive* study to prove that this material reacts with no other substance.

Defuse/Diffuse

- *Defuse = make harmless*
  The squad had to defuse the bomb within thirty minutes.
- *Diffuse = scattered*
  When the light passed through the special material, it became *diffused*.

Deprecate/Depreciate

- *Deprecate = deplore*
  Everybody *deprecated* the unruly actions of the striking workers in the factory.

- *Depreciation = reduction in value over time*
  The high *depreciation* charge of the manufacturing facility was a major reason for the company's fall in revenue this year.

Discrete/Discreet

- *Discrete = finite*
  Each *discrete* variable can take values from 10 through 20.
- *Discreet = confidential, careful without offending others*
  The manager made some *discreet* enquiries about the background of the candidate.

Edible/Eatable

- *Edible = fit to be eaten*
  All types of salts are not *edible*.
- *Eatables = food items*
  The employees went to get some *eatables* from the cafeteria.

Effluent/Affluent

- *Effluent = liquid waste*
  The manager of the chemical factory was trying hard to find a way of safely disposing off the *effluent*.
- *Affluent = wealthy*
  There are few *affluent* people who spend money for the betterment of those who are less fortunate.

Exhaustive/Exhausting

- *Exhaustive = comprehensive*
  We made an *exhaustive* study of the market position.
- *Exhausting = extremely tiring*
  The long and complicated experiment was *exhausting* for all scientists.

Extrapolate/Interpolate

- *Extrapolate = infer, extend the value of a curve*
  If we *extrapolate* the data, we can forecast the revenues for 2015.

- *Interpolate = placement between known values*
  The best way to estimate the missing data from 1996 through 1998 is to take the graph for the entire data and use mathematical *interpolation.*

Facilitate/Felicitate

- *Facilitate = to make something easier*
  It is hoped that the new methods adopted will *facilitate* better handling of the machinery.
- *Felicitate = to congratulate*
  The managing director *felicitated* the engineers of the company.

Farther/Further

- *Farther = great distance*
  I am so tired that I cannot go any *farther.*
- *Further = in addition*
  The chairperson *further* said that the goal of the company was to capture the market.

Fatal/Fateful

- *Fatal = leading to death or destruction*
  While attending the machines, the worker had a fatal fall in the factory.
- *Fateful = momentous*
  After that *fateful* decision by the government, the industry had to completely restructure itself.

Fewer/Less

- *Fewer = lesser number*
  The number of employees present at the meeting was *fewer* than what the manager had expected.
- *Less = lesser quantity*
  The quantity of coffee supplied to the company cafeteria was *less* than the stipulated amount.

Flair/Flare

- *Flair = talent*
  She was the only engineer in the class who had a *flair* for music.
- *Flared = emotional outburst*
  The scientist *flared* up when he was asked some embarrassing questions.

Flout/Flaunt

- *Flout = ignore the law/rule*
  They *flouted* the rules of the company.
- *Flaunt = show off*
  Parties present opportunities for people to flaunt their wealth.

Graceful/Gracious

- *Graceful = elegant, refined*
  The present CEO is known for tact and *graceful* behaviour.
- *Gracious = courteously*
  The hosting company treated all its guests *graciously*.

Horde/Hoard

- *Horde = a pack of*
  Hordes of animals charged towards the hunting party.
- *Hoard = collecting something, usually secretly*
  Hoarding essential commodities is a crime.

Ingenious/Ingenuous

- *Ingenious = creative, imaginative*
  The director highly commended the *ingenious* way of solving the mechanical problem evolved by the young engineer.
- *Ingenuous = candid, honest*
  The mechanic was pardoned for his frank and *ingenuous* admission of guilt.

Inimitable/Inimical

- *Inimitable = impossible to imitate*
  Everybody appreciated it when the chairman handled the client in his *inimitable* manner.

- *Inimical = hostile*
  In prehistoric days, humans had to contend with many *inimical* forces.

Inquiry/Enquiry

- *Inquiry = formal investigation*
  A security *inquiry* into the misappropriation of funds is going on.
- *Enquiry = to ask*
  The customer *enquired* about the availability of components from the visiting salesperson.

Luxuriant/Luxurious

- *Luxuriant = lavish, plentiful*
  The *luxuriant* growth of grass on the company's lawns is really eye soothing.
- *Luxurious = lavish, grand*
  A *luxurious* hotel was booked for the annual company party.

Masterly/Masterful

- *Masterly = brilliant*
  The *masterly* strokes of the strategist impressed everybody.
- *Masterful = expert*
  The *masterful* analysis of the market by the candidate impressed all.

Morale/Moral

- *Morale = emotional feeling*
  The *morale* of the team was high before the final game.
- *Moral = appropriate attitude, based on conscience*
  I can certify that the candidate bears a good *moral* character.

Official/Officious

- *Official = formal*
  I have received an *official* letter from the chairman.
- *Officious = bossy*
  Subordinates generally dislike an *officious* leader.

Ordinance/Ordnance

- *Ordinance = decree, order*
  The last government *ordinance* had a negative impact on the company's financials.
- *Ordnance = weapons*
  The *ordnance* factory in Kolkata is one of the oldest artillery factories in the country.

Perfunctory/Peremptory

- *Perfunctory = routine, superficial*
  Before beginning the speech, the leader exchanged *perfunctory* greetings with all attendees.
- *Peremptory = expecting compliance*
  The chairperson gave *peremptory* orders to the guilty worker and asked him to leave the campus.

Perpetrate/Perpetuate

- *Perpetrate = responsible for something wrong*
  Stricter laws are needed to halt the *perpetration* of heinous crimes.
- *Perpetuate = lasting, making something for a long time*
  A statue will be placed here to *perpetuate* the memory of the founder and chairman of the company.

Perspective/Prospective

- *Perspective = viewpoint*
  The director wanted to have a clear *perspective* of all the problems before taking any decision.
- *Prospective = potential, likely*
  The *prospective* manager was the main subject of discussion in the canteen, where all the employees met together.

Poignant/Pungent

- *Poignant = heartrending*
  The *poignant* analysis of the situation by the legal counsel moved everybody.

- *Pungent = strong, spicy*
  The *pungent* smell of the acid irritated me as soon as I entered the laboratory.

Pore/Pour

- *Pore = small opening*
  There are several tiny *pores* on the human skin.
- *Pour = discharge, flow*
  The technician *poured* the acid into the test-tube.

Prostrate/Prostate

- *Prostrate = in worship, lie with face down*
  They lay *prostrate* before the deity as a sign of devotion.
- *Prostate (gland) = male organ near the bladder*
  He has been hospitalized for the enlargement of his *prostate* glands.

Raise/Rise

- *Raise = increment*
  The workers demanded a *raise* in their salary.
- *Raise = intensify*
  Please *raise* the level of your performance.
- *Rise = growth*
  The *rise* in the price level was a concern for everybody.

Regal/Royal

- *Regal = imperial*
  Regal attire and behaviour are inappropriate for charity events.
- *Royal = related to the monarch*
  All members of the *royal* family enjoy certain privileges.
- *Royal = grand*
  The science conference was followed by a *royal* banquet dinner.

Site/Cite

- *Site = location, land*
  The chairman personally inspected the proposed *site* of the new building.

- *Cite = to support an argument*
  Please *cite* some examples to prove your point.

Skeptic/Septic

- *Skeptic = cynic*
  That man is really a *skeptic*; he does not even believe the detailed results.
- *Septic = infected*
  It is dangerous when a wound turns *septic*.

Some time/Sometime/Sometimes

- *Some time = a compound word meaning an indefinite period of time*
  The scientist spent some time analysing the results.
- *Sometime = an indefinite adverb meaning at one time or another*
  I hope to make it to the management cadre *sometime*.
- *Sometimes = an indefinite adverb meaning at different times*
  Sometimes the experiment may give unpredictable results.

Suite/Suit

- *Suite = set of rooms, collection or group*
  A beautiful *suite* in the best hotel of the city was kept reserved for the distinguished guest.
- *Suit = be right, clothes of same material*
  The climate of that city does not *suit* me.

Temerity/Timidity

- *Temerity = audacity, boldness*
  The *temerity* of the man when he demanded an immediate promotion surprised everybody.
- *Timidity = unassertive*
  You must shun *timidity* if you want to succeed in your ventures.

Troupe/Troop

- *Troupe = a group of entertainers, especially those who travel around*
  The performance of the dance *troupe* was highly applauded.

- *Troop = military unit, crowd*
The Indian *troop* easily repelled the intruders who had crossed the border.

Urbane/Urban

- *Urbane = suave, polished*
The *urbane* manners of the sophisticated person charmed everybody.
- *Urban = city, town*
The teledensity in the *urban* population is over 40 per cent.

Voucher/Coupon

- *Voucher = documentary evidence, receipt proving a purchase*
He requested for a *voucher* against the purchases.
- *Coupons = token, slip*
At the end of the function, *coupons* were given to the employees to enable them to have lunch in the company canteen.

Appendix C

Technical Guidelines for IEEE, ACM, and VDAT Papers

It is critical for engineers and scientists to keep an open and continuous communication channel with their peer community. The purpose and mechanism of the exchanges may differ. However, the intent is always the same: to exchange the most recent findings with peers around the globe. In order to facilitate a common understanding amongst all scientists and engineers worldwide, it is imperative that scientists and engineers write according to a specified format. Most reputed international organizations, such as the IEEE, ACM, etc., have clear style guides for submission and printing of technical papers. The guidelines from IEEE, ACM, and VDAT have been included below, for ready reference.

Preparation of Papers for IEEE TRANSACTIONS and JOURNALS (June 2003)

First A. Author, Second B. Author, Jr., and Third C. Author, *Member, IEEE*

Abstract—These instructions give you guidelines for preparing papers for IEEE transactions and journals. Use this document as a template if you are using Microsoft *Word* 6.0 or later. Otherwise, use this document as an instruction set. The electronic file of your paper will be formatted further at IEEE. Define all symbols used in the abstract. Do not cite references in the abstract. Do not delete the blank line immediately above the abstract; it sets the footnote at the bottom of this column.

Index Terms—About four key words or phrases in alphabetical order, separated by commas. For a list of suggested keywords, send a blank e-mail to keywords@ieee.org or visit the IEEE web site at http://www.ieee.org/web/developers/webthes/index.htm.

1. Introduction

This document is a template for Microsoft Word versions 6.0 or later. If you are reading a paper version of this document, please download the electronic file, TRANS-JOUR.DOC, from http://www.ieee.org/organizations/pubs/transactions/stylesheets.htm so you can use it to prepare your manuscript. If you would prefer to use LATEX, download IEEE's LATEX style and sample files from the same Web page. Use these LATEX files for formatting, but please follow the instructions in TRANS-JOUR.DOC or TRANS-JOUR.PDF.

If your paper is intended for a conference, please contact your conference editor concerning acceptable word processor formats for your particular conference.

When you open TRANS-JOUR.DOC, select 'Page Layout' from the 'View' menu in the menu bar, which allows you to see the footnotes. Then type over sections of TRANS-JOUR.DOC or cut and paste from another document and then use markup styles. The pull-down style menu is at the left of the Formatting Toolbar at the top of your *Word* window (for example, the style at this point in the document is 'Text'). Highlight a section that you want to designate with a certain style, then select the appropriate name on the style menu. The style will adjust your fonts and line spacing. *Do not change the font sizes or line spacing to squeeze more text into a limited number of pages.* Use italics for emphasis; do not underline.

To insert images in *Word,* position the cursor at the insertion point and either use Insert | Picture | From File or copy the image to the Windows clipboard and then Edit | Paste Special | Picture (with 'Float over text' unchecked).

IEEE will do the final formatting of your paper. If your paper is intended for a conference, please observe the conference page limits.

2. Procedure for Paper Submission

Review Stage

Please check with your editor on whether to submit your manuscript by hard copy or electronically for review. If hard copy, submit photocopies such that only one column appears per page. This will give your referees plenty of room to write comments. Send the number of copies specified by your editor (typically four). If submitted electronically, find out if your editor prefers submissions on disk or as e-mail attachments.

If you want to submit your file with one column electronically, please do the following:

- First, click on the View menu and choose Print Layout.
- Second, place your cursor in the first paragraph. Go to the Format menu, choose Columns, choose one column Layout, and choose 'apply to whole document' from the dropdown menu.
- Third, click and drag the right margin bar to just over 4 inches in width.

The graphics will stay in the 'second' column, but you can drag them to the first column. Make the graphic wider to push out any text that may try to fill in next to the graphic.

Final Stage

When you submit your final version, after your paper has been accepted, print it in two-column format, including figures and tables. Send three prints of the paper; two will go to IEEE and one will be retained by the Editor-in-Chief or conference publications chair.

You must also send your final manuscript on a disk, which IEEE will use to prepare your paper for publication. Write the authors' names on the disk label. If you are using a Macintosh, please save your file on a PC formatted disk, if possible. You may use *Zip* or CD-ROM disks for large files, or compress files using *Compress, Pkzip, Stuffit,* or *Gzip.*

Also send a sheet of paper with complete contact information for all authors. Include full mailing addresses, telephone numbers, fax numbers, and e-mail addresses. This information will be used to send each author a complimentary copy of the journal in which the paper appears. In addition, designate one author as the 'corresponding author.' This is the author to whom proofs of the paper will be sent. Proofs are sent to the corresponding author only.

Figures

All tables and figures will be processed as images. *However, IEEE cannot extract the tables and figures embedded in your document.* (The figures and tables you insert in your document are only to help you gauge the size of your paper, for the convenience of the referees, and to make it easy for you to distribute preprints.) Therefore, *submit, on separate sheets of paper, enlarged versions of the tables and figures that appear in your document.* These are the images IEEE will scan and publish with your paper.

Electronic Image Files (Optional)

You will have the greatest control over the appearance of your figures if you are able to prepare electronic image files. If you do not have the required computer skills, just submit paper prints as described above and skip this section.

1. *Easiest Way:* If you have a scanner, the best and quickest way to prepare noncolour figure files is to print your tables and figures on paper exactly as you want them to appear, scan them, and then save them to a file in PostScript (PS) or Encapsulated PostScript (EPS) formats. Use a separate file for each image. File names should be of the form 'fig1.ps' or 'fig2.eps'.

2. *Slightly Harder Way:* Using a scanner as above, save the images in TIFF format. High-contrast line figures and tables should be prepared with 600 dpi resolution and saved with no compression, 1 bit per pixel (monochrome), with file names of the form 'fig3.tif' or 'table1.tif'. To obtain a 3.45-in figure (one-column width) at 600 dpi, the figure requires a horizontal size of 2070 pixels. Typical file sizes will be on the order of 0.5 MB.

Photographs and grayscale figures should be prepared with 220 dpi resolution and saved with no compression, 8 bits per pixel (grayscale).

To obtain a 3.45-in figure (one-column width) at 220 dpi, the figure should have a horizontal size of 759 pixels.

Colour figures should be prepared with 400 dpi resolution and saved with no compression, 8 bits per pixel (palette or 256 colour). To obtain a 3.45-in figure (one column width) at 400 dpi, the figure should have a horizontal size of 1380 pixels.

For more information on TIFF files, please go to http://www.ieee.org/ organizations/pubs/transactions/information.htm and click on the link 'Guidelines for Author Supplied Electronic Text and Graphics'.

3. *Somewhat Harder Way:* If you do not have a scanner, you may create noncolour PostScript figures by 'printing' them to files. First, download a PostScript printer driver from http://www.adobe.com/support/downloads/pdrvwin.htm (for Windows) or from http://www.adobe.com/support/downloads/ pdrvmac.htm (for Macintosh) and install the 'Generic PostScript Printer' definition. In *Word,* paste your figure into a new document. Print to a file using the PostScript printer driver. File names should be of the form 'fig5.ps'. Use Adobe Type 1 fonts when creating your figures, if possible.

4. *Other Ways:* Experienced computer users can convert figures and tables from their original format to TIFF. Some useful image converters are Adobe *Photoshop,* Corel *Draw,* and Microsoft *Photo Editor,* an application that is part of Microsoft *Office 97* and *Office 2000* (look for C:\Program Files\Common Files \Microsoft Shared\ PhotoEd\ PHOTOED.EXE. (You may have to custom-install *Photo Editor* from your original *Office* disk.)

Here is a way to make TIFF image files of tables. First, create your table in *Word.* Use horizontal lines but no vertical lines. Hide gridlines (Table | Hide Gridlines). Spell check the table to remove any red underlines that indicate spelling errors. Adjust magnification (View | Zoom) such that you can view the entire table *at maximum area* when you select View | Full Screen. Move the cursor so that it is out of the way. Press 'Print Screen' on your keyboard; this copies the screen image to the Windows clipboard. Open Microsoft *Photo Editor* and click Edit | Paste as New Image. Crop the table image (click Select button; select

the part you want, then Image | Crop). Adjust the properties of the image (File | Properties) to monochrome (1 bit) and 600 pixels per inch. Resize the image (Image | Resize) to a width of 3.45 inches. Save the file (File | Save As) in TIFF with no compression (click 'More' button).

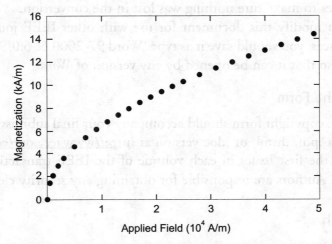

Fig. 1: Magnetization as a function of applied field. (Note that 'Fig.' is abbreviated. There is a period after the figure number, followed by two spaces. It is good practice to explain the significance of the figure in the caption.)

Most graphing programs allow you to save graphs in TIFF; however, you often have no control over compression or number of bits per pixel. You should open these image files in a program such as Microsoft *Photo Editor* and re-save them using no compression, either 1 or 8 bits, and either 600 or 220 dpi resolution (File | Properties; Image | Resize). See Section II-D2 for an explanation of number of bits and resolution. If your graphing program cannot export to TIFF, you can use the same technique described for tables in the previous paragraph.

A way to convert a figure from Windows Metafile (WMF) to TIFF is to paste it into Microsoft *PowerPoint*, save it in JPG format, open it with Microsoft *Photo Editor* or similar converter, and re-save it as TIFF.

Microsoft *Excel* allows you to save spreadsheet charts in Graphics Interchange Format (GIF). To get good resolution, make the *Excel* charts *very* large. Then use the 'Save as HTML' feature (see http://support.microsoft.com/support/ kb/articles/q158/0/79.asp). You can then convert from GIF to TIFF using Microsoft *Photo Editor,* for example.

No matter how you convert your images, it is a good idea to print the TIFF files to make sure nothing was lost in the conversion.

If you modify this document for use with other IEEE journals or conferences, you should save it as type 'Word 97-2000 & 6.0/95 - RTF (*.doc)' so that it can be opened by any version of *Word.*

Copyright Form

An IEEE copyright form should accompany your final submission. You can get a .pdf, .html, or .doc version at http://www.ieee.org/copyright or from the first issues in each volume of the IEEE transactions and journals. Authors are responsible for obtaining any security clearances.

3. Math

If you are using *Word,* use either the Microsoft Equation Editor or the *MathType* add-on (*http://www.mathtype.com*) for equations in your paper (Insert | Object | Create New | Microsoft Equation *or* MathType Equation). 'Float over text' should *not* be selected.

4. Units

Use either SI (MKS) or CGS as primary units. (SI units are strongly encouraged.) English units may be used as secondary units (in parentheses). *This applies to papers in data storage.* For example, write, '15 Gb/cm^2 (100 Gb/in^2)'. An exception is when English units are used as identifiers in trade, such as '3½ in disk drive'. Avoid combining SI and CGS units, such as current in amperes and magnetic field in oersteds. This often leads to confusion because equations do not balance dimensionally. If you must use mixed units, clearly state the units for each quantity in an equation.

The SI unit for magnetic field strength H is A/m. However, if you wish to use units of T, either refer to magnetic flux density B or

Table 1: Units for Magnetic Properties

Symbol	Quantity	Conversion from Gaussian and CGS EMU to SI[a]
Φ	magnetic flux	1 Mx $\rightarrow$ 10^{-8} Wb = 10^{-8} V·s
B	magnetic flux density, magnetic induction	1 G $\rightarrow$ 10^{-4} T = 10^{-4} Wb/m^2
H	magnetic field strength	1 Oe $\rightarrow$ $10^3/(4\pi)$ A/m
m	magnetic moment	1 erg/G = 1 emu $\rightarrow 10^{-3}$ A·m^2 = 10^{-3} J/T
M	magnetization	1 erg/(G·cm^3) = 1 emu/cm^3 $\rightarrow 10^3$ A/m
$4\pi M$	magnetization	1 G $\rightarrow$ $10^3/(4\pi)$ A/m
σ	specific magnetization	1 erg/(G·g) = 1 emu/g $\rightarrow$ 1 A·m^2/kg
j	magnetic dipole moment	1 erg/G = 1 emu $\rightarrow 4\pi \times 10^{-10}$ Wb·m
J	magnetic polarization	1 erg/(G·cm^3) = 1 emu/cm^3 $\rightarrow 4\pi \times 10^{-4}$ T
χ, κ	susceptibility	1 $\rightarrow 4\pi$
χ_ρ	mass susceptibility	1 cm^3/g $\rightarrow 4\pi \times 10^{-3}$ m^3/kg
μ	permeability	1 $\rightarrow 4\pi \times 10^{-7}$ H/m = $4\pi \times 10^{-7}$ Wb/(A·m)
μ_r	relative permeability	$\mu \rightarrow \mu_r$
w, W	energy density	1 erg/cm^3 $\rightarrow 10^{-1}$ J/m^3
N, D	demagnetizing factor	1 $\rightarrow 1/(4\pi)$

No vertical lines in table. Statements that serve as captions for the entire table do not need footnote letters.

[a]Gaussian units are the same as cgs emu for magnetostatics; Mx = maxwell, G = gauss, Oe = oersted; Wb = weber, V = volt, s = second, T = tesla, m = meter, A = ampere, J = joule, kg = kilogram, H = henry. magnetic field strength symbolized as $\mu_0 H$. Use the center dot to separate compound units, e.g., 'A·m^2'.

5. Helpful Hints

Figures and Tables

Because IEEE will do the final formatting of your paper, you do not need to position figures and tables at the top and bottom of each column. In fact, all figures, figure captions, and tables can be at the end of the paper. Large figures and tables may span both columns. Place figure captions below the figures; place table titles above the tables. If your figure has two parts, include the labels '(a)' and '(b)' as part of the artwork. Please verify that the figures and tables you mention in the text actually exist. *Please do not include captions as part of the figures. Do not put captions in 'text boxes' linked to the figures. Do not put borders around the outside of your figures.* Use the abbreviation 'Fig.' even at the beginning of a sentence. Do not abbreviate 'Table'. Tables are numbered with Roman numerals.

Colour printing of figures is available, but is billed to the authors (approximately $1,300, depending on the number of figures and number of pages containing colour). Include a note with your final paper indicating that you request colour printing. *Do not use colour unless it is necessary for the proper interpretation of your figures.* If you want reprints of your colour article, the reprint order should be submitted promptly. There is an additional charge of $81 per 100 for colour reprints.

Figure axis labels are often a source of confusion. Use words rather than symbols. As an example, write the quantity 'Magnetization', or 'Magnetization M', not just 'M'. Put units in parentheses. Do not label axes only with units. As in Fig. 1, for example, write 'Magnetization (A/m)' or 'Magnetization (A m-1)', not just 'A/m'. Do not label axes with a ratio of quantities and units. For example, write 'Temperature (K)', not 'Temperature/K'.

Multipliers can be especially confusing. Write 'Magnetization (kA/m)' or 'Magnetization (103 A/m)'. Do not write 'Magnetization (A/m) 1000' because the reader would not know whether the top axis label in Fig. 1 meant 16000 A/m or 0.016 A/m. Figure labels should be legible and of approximately 8 to 12 point type.

References

Number citations consecutively in square brackets [1]. The sentence punctuation follows the brackets [2]. Multiple references [2], [3] are each numbered with separate brackets [1]–[3]. When citing a section in a book, please give the relevant page numbers [2]. In sentences, refer simply to the reference number, as in [3]. Do not use 'Ref. [3]' or 'reference [3]' except at the beginning of a sentence: 'Reference [3] shows ...'. Unfortunately the IEEE document translator cannot handle automatic endnotes in *Word*; therefore, type the reference list at the end of the paper using the 'References' style.

Number footnotes separately in superscripts (Insert | Footnote)[1]. Place the actual footnote at the bottom of the column in which it is cited; do not put footnotes in the reference list (endnotes). Use letters for table footnotes (see Table 1).

Please note that the references at the end of this document are in the preferred referencing style. Give all authors names; do not use '*et al.*' unless there are six authors or more. Use a space after authors' initials. Papers that have not been published should be cited as 'unpublished' [4]. Papers that have been submitted for publication should be cited as 'submitted for publication' [5]. Papers that have been accepted for publication, but not yet specified for an issue should be cited as 'to be published' [6]. Please give affiliations and addresses for private communications [7].

Capitalize only the first word in a paper title, except for proper nouns and element symbols. If you are short of space, you may omit paper titles. However, paper titles are helpful to your readers and are strongly recommended. For papers published in translation journals, please give the English citation first, followed by the original foreign-language citation [8].

Abbreviations and Acronyms

Define abbreviations and acronyms the first time they are used in the text, even after they have already been defined in the abstract. Abbreviations such as IEEE, SI, ac, and dc do not have to be defined. Abbreviations that incorporate periods should not have spaces: write 'C.N.R.S.', not 'C. N. R. S'. Do not use abbreviations in the title unless they are unavoidable (for example, 'IEEE' in the title of this article).

---

[1] It is recommended that footnotes be avoided (except for the unnumbered footnote with the receipt date on the first page). Instead, try to integrate the footnote information into the text.

Equations

Number equations consecutively with equation numbers in parentheses flush with the right margin, as in (1). First use the equation editor to create the equation. Then select the 'Equation' markup style. Press the tab key and write the equation number in parenthesis. To make your equations more compact, you may use the solidus (/), the exp function, or appropriate exponents. Use parentheses to avoid ambiguities in denominators. Punctuate equations when they are part of a sentence, as in

$$\int_0^{r_2} F(r, \varphi)\, dr\, d\varphi = \left[\sigma r_2/(2\mu_0)\right] \tag{1}$$

$$\cdot \int_0^{\infty} \exp\left(-\lambda\left|z_j - z_i\right|\right) \lambda^{-1}\, J_1\left(\lambda r_2\right)\, J_0\left(\lambda r_i\right) d\lambda.$$

Be sure that the symbols in your equation have been defined before the equation appears or immediately following. Italicize symbols (T might refer to temperature, but T is the unit tesla). Refer to '(1)', not 'Eq. (1)' or 'equation (1)', except at the beginning of a sentence: 'Equation (1) is ...'.

Other Recommendations

Use one space after periods and colons. Hyphenate complex modifiers: 'zero-field-cooled magnetization'. Avoid danging participles, such as, 'Using (1), the potential was calculated'. It is not clear who or what used (1). Write instead, 'The potential was calculated by using (1)', or 'Using (1), we calculated the potential'.

Use a zero before decimal points: '0.25', not '.25'. Use 'cm^3', not 'cc'. Indicate sample dimensions as '0.1 cm × 0.2 cm', not '0.1 × 0.2 cm^2'. The abbreviation for 'seconds' is 's', not 'sec'. Do not mix complete spellings and abbreviations of units: use 'Wb/m^2' or 'webers per square meter', not 'webers/m^2'. When expressing a range of values, write '7 to 9' or '7-9', not '7~9'.

A parenthetical statement at the end of a sentence is punctuated outside of the closing parenthesis (like this). (A parenthetical sentence is punctuated within the parentheses.) In American English, periods and commas are within quotation marks, like 'this period'. Other punctuation is 'outside'! Avoid contractions; for example, write 'do not' instead of 'don't'. The serial comma is preferred: 'A, B, and C' instead of 'A, B and C'.

If you wish, you may write in the first person singular or plural and use the active voice ('I observed that ...' or 'We observed that ...' instead of 'It was observed that ...'). Remember to check spelling. If your native language is not English, please get a native English-speaking colleague to proofread your paper.

6. Some Common Mistakes

The word 'data' is plural, not singular. The subscript for the permeability of vacuum μ_0 is zero, not a lowercase letter 'o'. The term for residual magnetization is 'remanence'; the adjective is 'remanent'; do not write 'remnance' or 'remnant'. Use the word 'micrometer' instead of 'micron'. A graph within a graph is an 'inset', not an 'insert'. The word 'alternatively' is preferred to the word 'alternately' (unless you really mean something that alternates). Use the word 'whereas' instead of 'while' (unless you are referring to simultaneous events). Do not use the word 'essentially' to mean 'approximately' or 'effectively'. Do not use the word 'issue' as a euphemism for 'problem'. When compositions are not specified, separate chemical symbols by en-dashes; for example, 'NiMn' indicates the intermetallic compound $Ni_{0.5} Mn_{0.5}$ whereas 'Ni–Mn' indicates an alloy of some composition $Ni_x Mn_{1-x}$.

Be aware of the different meanings of the homophones 'affect' (usually a verb) and 'effect' (usually a noun), 'complement' and 'compliment', 'discreet' and 'discrete', 'principal' and 'principle', etc. Do not confuse 'imply' and 'infer'.

Prefixes such as 'non', 'sub', 'micro', 'multi', and 'ultra' are not independent words; they should be joined to the words they modify, usually without a hyphen. There is no period after the 'et' in the Latin abbreviation *et al.* (it is also italicized). The abbreviation 'i.e.' means 'that is' and the abbreviation 'e.g.' means 'for example', (these abbreviations are not italicized).

An excellent style manual and source of information for science writers is [9]. A general IEEE style guide, *Information for Authors,* is available at http://www.ieee.org/organizations/pubs/transactions/information.htm

7. Editorial Policy

Submission of a manuscript is not required for participation in a conference. Do not submit a reworked version of a paper you have submitted or published elsewhere. Do not publish 'preliminary' data or results. The submitting author is responsible for obtaining the agreement of all coauthors and any consent required from sponsors before submitting a paper. IEEE transactions and journals strongly discourage courtesy authorship. It is the obligation of the authors to cite relevant prior work.

The Transactions and Journals Department does not publish conference records or proceedings. The transactions does publish papers related to conferences that have been recommended for publication on the basis of peer review. As a matter of convenience and service to the technical community, these topical papers are collected and published in one issue of the transactions.

At least two reviews are required for every paper submitted. For conference-related papers, the decision to accept or reject a paper is made by the conference editors and publications committee; the recommendations of the referees are advisory only. Undecipherable English is a valid reason for rejection. Authors of rejected papers may revise and resubmit them to the transactions as regular papers, whereupon they will be reviewed by two new referees.

8. Publication Principles

The contents of IEEE transactions and journals are peer-reviewed and archived. The transactions publishes scholarly articles of archival value as well as tutorial expositions and critical reviews of classical subjects and topics of current interest.

Authors should consider the following points:

1. Technical papers submitted for publication must advance the state of knowledge and must cite relevant prior work.
2. The length of a submitted paper should be commensurate with the importance, or appropriate to the complexity, of the work. For example, an obvious extension of previously published work might not be appropriate for publication or might be adequately treated in just a few pages.
3. Authors must convince both peer reviewers and the editors of the scientific and technical merit of a paper; the standards of proof are higher when extraordinary or unexpected results are reported.
4. Because replication is required for scientific progress, papers submitted for publication must provide sufficient information to allow readers to perform similar experiments or calculations and use the reported results. Although not everything need be disclosed, a paper must contain new, useable, and fully described information. For example, a specimen's chemical composition need not be reported if the main purpose of a paper is to introduce a new measurement technique. Authors should expect to be challenged by reviewers if the results are not supported by adequate data and critical details.
5. Papers that describe ongoing work or announce the latest technical achievement, which are suitable for presentation at a professional conference, may not be appropriate for publication in a transactions or journal.

9. Conclusion

A conclusion section is not required. Although a conclusion may review the main points of the paper, do not replicate the abstract as the conclusion. A conclusion might elaborate on the importance of the work or suggest applications and extensions.

Appendix

Appendixes, if needed, appear before the acknowledgment.

Acknowledgment

The preferred spelling of the word 'acknowledgment' in American English is without an 'e' after the 'g'. Use the singular heading even if you have many acknowledgments. Avoid expressions such as, 'One of us (S.B.A.) would like to thank ...'. Instead, write 'F. A. Author thanks ...'. *Sponsor and financial support acknowledgments are placed in the unnumbered footnote on the first page.*

References

1. G. O. Young, "Synthetic structure of industrial plastics (Book style with paper title and editor)," in *Plastics*, 2nd ed. vol. 3, J. Peters, Ed. New York: McGraw-Hill, 1964, pp. 15–64.

2. W.-K. Chen, *Linear Networks and Systems* (Book style). Belmont, CA: Wadsworth, 1993, pp. 123–135.

3. H. Poor, *An Introduction to Signal Detection and Estimation*. New York: Springer-Verlag, 1985, ch. 4.

4. B. Smith, 'An approach to graphs of linear forms (Unpublished work style)," unpublished.

5. E. H. Miller, "A note on reflector arrays (Periodical style—Accepted for publication)," *IEEE Trans. Antennas Propagat.*, to be published.

6. J. Wang, "Fundamentals of erbium-doped fiber amplifiers arrays (Periodical style—Submitted for publication)," *IEEE J. Quantum Electron.*, submitted for publication.

7. C. J. Kaufman, Rocky Mountain Research Lab., Boulder, CO, private communication, May 1995.

8. Y. Yorozu, M. Hirano, K. Oka, and Y. Tagawa, "Electron spectroscopy studies on magneto-optical media and plastic substrate interfaces(Translation Journals style)," *IEEE Transl. J. Magn.Jpn.*, vol. 2, Aug. 1987, pp. 740–741 [*Dig. 9th Annu. Conf. Magnetics* Japan, 1982, p. 301].

9. M. Young, *The Techincal Writers Handbook*. Mill Valley, CA: University Science, 1989.

10. J. U. Duncombe, "Infrared navigation—Part I: An assessment of feasibility (Periodical style)," *IEEE Trans. Electron Devices*, vol. ED-11, pp. 34–39, Jan. 1959.

11. S. Chen, B. Mulgrew, and P. M. Grant, "A clustering technique for digital communications channel equalization using radial basis function networks," *IEEE Trans. Neural Networks*, vol. 4, pp. 570–578, July 1993.

12. R. W. Lucky, "Automatic equalization for digital communication," *Bell Syst. Tech. J.*, vol. 44, no. 4, pp. 547–588, Apr. 1965.

13. S. P. Bingulac, "On the compatibility of adaptive controllers (Published Conference Proceedings style)," in *Proc. 4th Annu. Allerton Conf. Circuits and Systems Theory*, New York, 1994, pp. 8–16.

14. G. R. Faulhaber, "Design of service systems with priority reservation," in *Conf. Rec. 1995 IEEE Int. Conf. Communications*, pp. 3–8.

15. W. D. Doyle, "Magnetization reversal in films with biaxial anisotropy," in *1987 Proc. INTERMAG Conf.*, pp. 2.2-1–2.2-6.

16. G. W. Juette and L. E. Zeffanella, "Radio noise currents n short sections on bundle conductors (Presented Conference Paper style)," presented at the IEEE Summer power Meeting, Dallas, TX, June 22–27, 1990, Paper 90 SM 690-0 PWRS.

17. J. G. Kreifeldt, "An analysis of surface-detected EMG as an amplitude-modulated noise," presented at the 1989 Int. Conf. Medicine and Biological Engineering, Chicago, IL.

18. J. Williams, "Narrow-band analyzer (Thesis or Dissertation style)," Ph.D. dissertation, Dept. Elect. Eng., Harvard Univ., Cambridge, MA, 1993.

19. N. Kawasaki, "Parametric study of thermal and chemical nonequilibrium nozzle flow," M.S. thesis, Dept. Electron. Eng., Osaka Univ., Osaka, Japan, 1993.

20. J. P. Wilkinson, "Nonlinear resonant circuit devices (Patent style)," U.S. Patent 3 624 12, July 16, 1990.

21. *IEEE Criteria for Class IE Electric Systems* (Standards style), IEEE Standard 308, 1969.

22. *Letter Symbols for Quantities*, ANSI Standard Y10.5-1968.

23. R. E. Haskell and C. T. Case, "Transient signal propagation in lossless isotropic plasmas (Report style)," USAF Cambridge Res. Lab., Cambridge, MA Rep. ARCRL-66-234 (II), 1994, vol. 2.

24. E. E. Reber, R. L. Michell, and C. J. Carter, "Oxygen absorption in the Earth's atmosphere," Aerospace Corp., Los Angeles, CA, Tech. Rep. TR-0200 (420-46)-3, Nov. 1988.

25. (Handbook style) *Transmission Systems for Communications,* 3rd ed., Western Electric Co., Winston-Salem, NC, 1985, pp. 44–60.

26. *Motorola Semiconductor Data Manual,* Motorola Semiconductor Products Inc., Phoenix, AZ, 1989.

27. (Basic Book/Monograph Online Sources) J. K. Author. (year, month, day). *Title* (edition) [Type of medium]. Volume(issue). Available: http://www.(URL)

28. J. Jones. (1991, May 10). Networks (2nd ed.) [Online]. Available: http://www.atm.com

29. (Journal Online Sources style) K. Author. (year, month). Title. *Journal* [Type of medium]. Volume(issue), paging if given. Available: http://www.(URL)

30. R. J. Vidmar. (1992, August). On the use of atmospheric plasmas as electromagnetic reflectors. *IEEE Trans. Plasma Sci.* [Online]. *21(3).* pp. 876—880. Available: http://www.halcyon.com/pub/journals/21ps03-vidmar

First A. Author (M'76–SM'81–F'87) and the other authors may include biographies at the end of regular papers. Biographies are often not included in conference-related papers. This author became a Member (M) of IEEE in 1976, a Senior Member (SM) in 1981, and a Fellow (F) in 1987. The first paragraph may contain a place and/or date of birth (list place, then date). Next, the author's educational background is listed. The degrees should be listed with type of degree in what field; which institution, city, state, or country; and the year degree was earned. The author's major field of study should be in lower-case.

The second paragraph uses the pronoun of the person (he or she) and not the author's last name. It lists military and work experience, including summer and fellowship jobs. Job titles are capitalized. The current job must have a location; previous positions may be listed without one. Information concerning previous publications may be included. Try not to list more than three books or published articles. The format for listing publishers of a book within the biography is: title of book (city, state: publishers name, year) similar to a reference. Current and previous research interests end the paragraph.

The third paragraph begins with the author's title and last name (e.g., Dr. Smith, Prof. Jones, Mr. Kajor, Ms. Hunter). List any memberships in professional societies other than the IEEE. Finally, list any awards and work for IEEE committees and publications. If a photograph is provided, the biography will be indented around it. The photograph is placed at the top left of the biography. Personal hobbies will be deleted from the biography.

ACM Word Template for SIG Site

1st author
1st author's affiliation
1st line of address
2nd line of address
Telephone number, incl. country code
1st author's e-mail address

2nd author
2nd author's affiliation
1st line of address
2nd line of address
Telephone number, incl. country code
2nd e-mail

3rd author
3rd author's affiliation
1st line of address
2nd line of address
Telephone number, incl. country code
3rd e-mail

Abstract

In this paper, we describe the formatting guidelines for ACM SIG Proceedings.

Categories and Subject Descriptors

D.3.3 [Programming Languages]: Language Contructs and Features— *abstract data types, polymorphism, control structures.* This is just an example, please use the correct category and subject descriptors for your submission. The ACM computing classification scheme: http://www.acm.org/class/1998/

General Terms

The general terms must be any of the following 16 designated terms: algorithms, management, measurement, documentation, performance, design, economics, reliability, experimentation, security, human factors, standardization, languages, theory, legal aspects, verification.

Keywords

Keywords are your own designated keywords.

10. Introduction

The proceedings are the records of the conference. ACM hopes to give these conference by-products a single and high-quality appearance. To do this, we ask that authors follow some simple guidelines. In essence, we ask you to make your paper look exactly like this document. The easiest way to do this is simply to down-load a template from [2], and replace the content with your own material.

11. Page Size

All material on each page should fit within a rectangle of 18 × 23.5 cm (7" × 9.25"), centered on the page, beginning 2.54 cm (1") from the top of the page and ending with 2.54 cm (1") from the bottom. The right and left margins should be 1.9 cm (.75"). The text should be in two 8.45 cm (3.33") columns with a .83 cm (.33") gutter.

12. Typeset Text

12.1 Normal or Body Text

Permission to make digital or hard copies of all or part of this work for personal or classroom use is granted without fee, provided that copies are not made or distributed for profit or commercial advantage and that copies bear this notice and the full citation on the first page. To copy otherwise, to republish, to post on servers, or to redistribute to lists, requires prior specific permission and/or a fee.

Conference'04, Month 1–2, 2004, City, State, Country.
Copyright 2004 ACM 1-58113-000-0/00/0004...$5.00.

Please use a 9-point Times Roman font, or some other Roman font with serifs, as close as possible in appearance to Times Roman. The goal is to have a 9-point text. Please use sans-serif or non-proportional fonts only for special purposes, such as distinguishing source code text. If Times Roman is not available, try the font named Computer Modern Roman. On a Macintosh, use the font named Times. Right margins should be justified, not ragged.

12.2 Title and Authors

The title (Helvetica 18-point bold), authors' names (Helvetica 12-point), and affiliations (Helvetica 10-point) run across the full width of the page—one column wide. We also recommend phone number (Helvetica 10-point) and e-mail address (Helvetica 12-point). If only one address is needed, center all address text. For two addresses, use two centered tabs, and so on. For more than three authors, you may have to improvise[2].

12.3 First Page Copyright Notice

Please leave 3.81 cm (1.5") of blank text box at the bottom of the left column of the first page for the copyright notice.

---

[2] If necessary, you may place some address information in a footnote, or in a named section at the end of your paper.

12.4 Subsequent Pages

For pages other than the first page, start at the top of the page, and continue in double-column format. The two columns on the last page should be as close to equal length as possible.

Table 1: Table captions should be placed above the table

Graphics	Top	In-between	Bottom
Tables	End	Last	First
Figures	Good	Similar	Very well

12.5 References and Citations

Footnotes should be Times New Roman 9-point, and justified to the full width of the column.

Use the standard Communications of the ACM format for references—that is, a numbered list at the end of the article, ordered alphabetically by first author, and referenced by numbers in brackets [1]. See the examples of citations at the end of this document. Within this template file, use the style named references for the text of your citation.

The references are also in 9 pt, but that section is ragged right. References should be published materials accessible to the public. Internal technical reports may be cited only if they are easily accessible (i.e., you can give the address to obtain the report within your citation) and may be obtained by any reader. Proprietary information may not be cited. Private communications should be acknowledged, not referenced (e.g., '[Robertson, personal communication]').

12.6 Page Numbering, Headers, and Footers

Do not include headers, footers, or page numbers in your submission. These will be added when the publications are assembled.

13. Figures/Captions

Place tables, figures, and images in text as close to the reference as possible (see Figure 1). It may extend across both columns to a maximum width of 17.78 cm (7").

Captions should be in Times New Roman 9-point bold. They should be numbered (e.g., 'Table 1' or 'Figure 2'), please note that the words table and figure are spelled out. Figure captions should be centered beneath the image or picture, whereas table captions should be centered above the table body.

14. Sections

The heading of a section should be in Times New Roman 12-point bold in all-capitals flush left with an additional 6-points of white space above the section head. Sections and subsequent sub-sections should be numbered and flush left. For a section head and a sub-section head together (such as Section 3 and sub-section 3.1), use no additional space above the sub-section head.

14.1 Sub-sections

The heading of sub-sections should be in Times New Roman 12-point bold with only the initial letters capitalized. (Note: For subsections and sub-sub-sections, a word like *the* or *a* is not capitalized unless it is the first word of the header.)

14.1.1 Sub-sub-sections

The heading for sub-sub-sections should be in Times New Roman 11-point italic with initial letters capitalized and 6-points of white space above the sub-sub-section head.

14.1.1.1 Sub-sub-sections

The heading for sub-sub-sections should be in Times New Roman 11-point italic with initial letters capitalized.

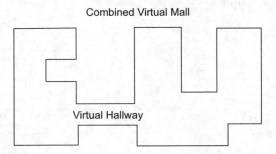

Fig. 1: Insert caption to place caption below figure.

14.1.1.2 Sub-sub-sections
The heading for sub-sub-sections should be in Times New Roman 11-point italic with initial letters capitalized.

15. Acknowledgments

Our thanks to ACM SIGCHI for allowing us to modify the templates they had developed.

16. References

1. Bowman, B., Debray, S. K., and Peterson, L. L. Reasoning about naming systems. *ACM Trans. Program. Lang. Syst., 15,* 5 (Nov. 1993), 795-825.
2. Ding, W., and Marchionini, G. *A Study on Video Browsing Strategies.* Technical Report UMIACS-TR-97-40, University of Maryland, College Park, MD, 1997.
3. Fröhlich, B. and Plate, J. The cubic mouse: a new device for three-dimensional iput. In *Proceedings of the SIGCHI conference on Human factors in computing systems (CHI '00)* (The Hague, The Netherlands, April 1-6, 2000). ACM Press, New York, NY, 2000, 526-531.
4. Lamport, L. *LaTeX User's Guide and Document Reference Manual.* Addison-Wesley, Reading, MA, 1986.
5. Sannella, M. J. *Constraint Satisfaction and Debugging for Interactive User Interfaces.* Ph.D. Thesis, University of Washington, Seattle, WA, 1994.

Title: How to Type-set your Paper for VDAT 2005 Proceedings

Elite Publishers[3]

Abstract

This document provides instructions to authors of papers accepted for presentation in the seventh VLSI Design and Test Symposium to be held in Bangalore during August 10–13, 2005. Please follow the following format guidelines carefully. Author(s) are required to compose the text, drawings, photographs, tables, and slides within the range of print area (4.5" × 8").

1. Page Limit

Regular papers and tutorials will have a limit of 10 pages, short papers will have a limit of 8 pages and poster papers will have a limit of 4 pages. The typesetting rules described here apply to regular papers, tutorials, short papers, and also to the poster papers. Authors have the option of buying at most, two extra pages by paying Rs. 1000/- per page. Send the cheque to Mr Gopal Naidu, Finance Chair, VDAT 2005, Texas Instruments (India) Pvt. Ltd, Bagmane Tech Park, Opposite LRDE, C. V. Raman Nagar Post, Bangalore - 560093; and make the draft payable to 'VLSI Design and Test Symposium 2005'.

2. Paper (Text Matter)

Final manuscripts must be submitted electronically at the VDAT 2005 website. Please do not send hard copies by regular mail or electronic copies by e-mail. Papers must be formatted using MS Word. The manuscript should be typewritten in single line spacing and standard font spacing, with 10 points font size in Times New Roman script.

---

[3] Contact Information: Elite Publishers, New Delhi. Rg_elite@yahoo.com

3. Presentation

Abstract, introduction, sub-headings, paragraphs, and conclusions of the manuscript should begin at the left margin and should be consistent throughout the paper.

4. Arrangement of the Manuscript

The manuscript should contain following points arranged in the order specified below:

4.1 Title

The Title of the paper should be in **bold capital letters** with center justification at the third line of the first page in 16-points size.

4.2 Name(s) of the Author(s), Title(s), and Affiliation

Typewrite the name of the author(s) in bold 12-points font size followed by their title(s). Leave one line space in between the title of the paper and the name of the author(s). Print affiliation of the Author(s) in 9 points front size in footnotes.

4.3 Abstract

Synopsis of not more than 200 words should be written on a new paragraph leaving one line space after the name(s) of the author(s). Use 10-points font size. The abstract should clearly focus on the scope of the paper and the main conclusions reached.

4.4 Keywords (Index)

List out key words in Italics and 10 points size in a separate sheet.

4.5 Tables and Figures

All graphic items must fit within the print area. Lettering within the tables and figures should be 10 points. Please do not use colour in your figures or any part of the manuscript. If you plan to include an image, ensure that it is of high quality so that when printed it will reproduce clearly. All figures and tables must have captions and must be numbered.

4.6 Sections

Number your sections 1, 2, 3, etc. and subsections 2.1, 3.2.1, etc.

4.7 References

Within the text, references should be indicated by citing the surname of the author(s) followed by the year of publication. The abbreviated author and date references should be placed in parentheses, e.g., Hasan and Khan (1998), or Agrawal (1997). Here is a sample reference:

Hasan, S. And Khan, M.A. (1998), Management of Calcutta Megacity: A transition from a reactive to a proactive strategy, In Sustainable Development, Vol. 6, No 2, August, pp. 55–67.

For all conventions, references should follow the Harvard system (e.g. the Chicago Manual of Style, Style B).

5. Slides

Do not use animation or colour filling in your slides. We plan to reproduce 2 slides per page. When printed, none of the lettering in your slides must be less than 10 point in size. We recommend that you use at least 18 point font size in your slides.

6. Copyrights

Any necessary rights or permission to reproduce quoted material or illustrations published elsewhere remain the responsibility of the author(s). To safeguard authors' rights, the copyright of all material published in the proceeding is vested in the publisher.

7. Submission

Submit the final camera-ready version electronically.

References

1. Hasan, S. And Khan, M.A. (1998), *Management of Calcutta Megacity: A transition from a reactive to a proactive strategy*, In *Sustainable Development*, Vol. 6, No 2, August, pp. 55–67.

Appendix A: Microsoft Word Setting

We recommend the following procedure when you use *Microsoft Word* to typeset your document. Go to File | Page | Setup menu and select paper size to be Letter (8.5" × 11"). Then set margins as follows—top and bottom margins of 1.5", and left and right margins of 2" each. Set gutter to 0" and header and footer margins to 0.5".

Guidelines for Presentations

A digital light processing projector, collar mike, and laser pointer will be available for the presentations as audio-visual aids. Presentation foils must adhere to the following guidelines:

1. for a 20-minute presentation, do not make more than 10 foils.
2. use '*VDAT 2005*' as a footer.
3. the normal flow of the presentation must be:
    *   title slide
    *   outline
    *   background and previous work
    *   your contribution
    *   experimental results
    *   summary
4. clearly indicate who the presenter is and who the coauthors are on the title foil.
5. highlight your contributions and compare your work with previous work—use the time judiciously.
6. do not spend too much time providing the background!
7. do not reuse material from any source without acknowledgement
8. do not use fonts less than 20 point
9. use Arial font
10. do not clutter your foils with information
11. do not cut and paste paragraphs from your paper
12. if animation is used, please ensure that the printed version will still make sense. We wish to make the PDF versions of the foils available on the CD.

13. also, please send a brief bio-data to <u>vdat05@hotmail.com</u> - the bio must include:
    - name of the presenter
    - session number (as indicated in the <u>Technical Program</u>)
    - introduction to the speaker—educational background, professional background, areas of interest, awards, etc. Please ensure this is brief. This material will be used to introduce the speaker by the session chair.

Submitting authors must become members of the <u>VDAT Mailing List</u>, where updates on the symposium will be sent.

Bibliography

21st Century Grammar Handbook, Dell Publishing, New York, 1993.

American National Standard for the Preparation of Scientific Papers for Written and Oral Presentation, American National Standards Institute, New York, 1979.

Angell, David and Brent Helper 1994, *Elements of Email Style*, Addison-Wesley Publishing Company, New York.

Barrass, Robert 1978, *Scientists Must Write: A Guide to Better Writing for Scientists, Engineers, and Students*, Chapman & Hall, New York.

Benhabib, S. 1992, *Situating the Self: Gender, Community and Postmodernism in Contemporary Ethics*, Cambridge.

Bennett, Millard and J.D.Corrigan 1981, *Successful Communication and Effective Speaking*, Parking Publishing Company Inc., Bombay.

Benveniste, E. 1973, *Indo-European Language and Society*, University of Miami Press, Miami.

Bernstein T. M. 1977, The *Careful Writer: A Modern Guide to English Usage*, Atheneum, New York.

Booth V.H. 1993, *Communicating in Science: Writing a Scientific Paper and Speaking at Scientific Meetings*, 2nd edn, Cambridge University Press.

Booth, V. 1981, *Writing a scientific paper and speaking at scientific meetings*, 5th edn, The Biochemical Society, London.

Briggs, A. and P. Burke 2002, *A Social History of Media: from Gutenberg to the Internet*, Cambridge, Polity

Briscoe, M. H. 1978, *Preparing scientific illustration Scientists, Engineers, and Students*, Chapman and Hall, New York.

Burch, G. E. 1954, *Of publishing scientific papers*, Grune and Stratton, New York.

Chatterjea, Gautam 2005, The *Art & Science of Presentations*, Rupa and Co.

Chicago Manual of Style, 14th edn, University of Chicago Press, 1993.

Comfort Jeremy et al. 1984, *Business Reports in English*, Cambridge University Press.

Cook, C. K. 1985, *Line by Line: How to Edit Your Own Writing*, Houghton Mifflin, Boston.

Day, Robert A. 1988, *How to Write and Publish a Scientific Paper*, 3rd edn, Oryx Press, Phoenix.

Day, Robert A. 1995, *Scientific English: A Guide for Scientists and Other Professionals*, 2nd edn, Oryx Press, Phoenix, Arizona.

Denzin, N. K. 1991, *Images of Postmodern Society*, Sage, Newbury Park CA.

Division of Humanities and Social Sciences, Anna University 1990, *English for Engineers and Technologists, Vol. I and II*, Orient Longman, Hyderabad,

Dubinko, Svetlana and Ludmila Koledenkova 2002, *Mastering Business English, Clarity in Business Expression*, Orient Longman Limited, Hyderabad.

Dugger, Jim 2000, *Business Letters for Busy People*, Jaico Publishing House, Mumbai.

Dutta, Banani and S.S. Bhattacharyya 1993, 'Two-photon dissociation of HD^+ in two-frequency laser fields in the presence of two intermediate resonances', J. Phys. B: At. Mol. Opt. Phys. 26, IOP Publishing Ltd.

Eisenberg, Anne 1989, *Writing Well for the Technical Professions*, Harper & Row, New York.

Ellul, J. 1980, *The Technological System*, Continuum, New York.

Elsenstein, E. 1979, *The Printing Press as an Agent of Change: Communication and Cultural Transformation in Early Modern Europe, 2 Vols.*, Cambridge University Press,

Flesch R.F. 1962, *The Art of Readable Writing*, 1st edn, Collier, New York.

Flynn, Nancy and Tim Flynn 1964, *Writing Effective E-mail*, Harper and Row, New York.

Fowler, H .W. 1975, *A Dictionary of Modern English Usage*, Oxford University Press, London.

Fowler, H.W. as cited in Gowers, Sir Ernest 1954, *The Complete Plain Words*, Penguin Books.

Godin, Seth 1993, *The Smiley Dictionary*, Peachpit,

Gunning, Robert 1959, *The technique of clear writing*, McGraw-Hill, New York.

Gustavi , Bjorn 2003, *How to Write and Illustrate a Scientific Paper*, Cambridge University Press.

Hartman, Diane B. and Karen Nantz 1996, *The three R's of E-mail: Rights and Responsibilities*, Crisp Publications.

Hebdige, D. 1979, *Subculture: The Meaning of Style*, Methuen, New York.

Heyman, Richard 1999, *Why Didn't You Say That in the First Place? How to Be Understood at Work,* Jossey-Bass Publishers, (place)

Houghton, B. 1971, *Scientific Periodicals, Their Historical Development, Characteristics and Control*, Shoestring Press, Hamden, CT.

http://en.wikipedia.org/wiki/Computer_virus 8th March 2006

http://www.cxotoday.com/cxo/jsp/article.jsp?article_id=68148&cat_id = 909, 8th March 2006.

http://www.timesonline.co.uk/article/0,,1-979473,00.html 8th March 2006

Humphreys, Gordon 1971, *English Grammar,* Teach Yourself Books, St. Paul's House, London.

Jones W.P. and Keene M.L. 1981, *Writing Scientific Papers and Reports.* 8th edn, WC Brown Co., Dubuque, Iowa

Kelly, R.A. 1970, *The Use of English for Technical Students*, Harrap and Co., London.

King, Lester S. 1991, *Why Not Say It Clearly: A Guide to Scientific Writing,* 2nd edn, Little, Brown: Boston, MA.

Kokrady, A. and C.P. Ravikumar 2003, 'Static verification of test vectors for IR drop failure', International Conference on Computer-Aided Design (ICCAD), November 9–13, San Jose, USA.

Krug, Gary 2005, *Communication, Technology and Cultural Change*, Sage Publication, London.

Kuhn, T. 1970, *The Structure of Scientific Revolutions*, 2nd edn, University of Chicago Press, Chicago.

Kumar, Rahul and C. P. Ravikumar 2002, 'Leakage Power Estimation for Deep Submicron Circuits in an ASIC Design Environment', Proceedings of the 2002 conference on Asia South Pacific design automation/VLSI Design, January 07–11.

Kumar, Ravindra G., et al. 1996 'Molecular pendular states in intense laser fields', *Physical Review*, vol. 53, issue 5.

Kumarappan, V., M. Krishnamurthy, D. Mathur, and L. C. Tribedi 2001, 'Effect of laser polarization on x-ray emission from Ar_n (n=200–10^4) clusters in intense laser fields', *Physical Review,* vol. 63, issue 2.

Lamport, L. 1994, *A Document Preparation System: User's Guide and Reference Manual*, 2nd edn, Addison-Wesley, Reading, MA.

Lederer, Richard 1987, *Anguished English*, Dell Publishers, New York.

Leigh, Andrew 2000, *Persuasive reports and proposals*, Universities Press India Ltd, Hyderabad.

Maggie, R. 1991, *The dictionary of bias-free usage: A Guide to Non-Discriminating Langauge*, Oryx Press, Phonenix.

Mandal, C.A., P.P. Chakrabarty, and S. Ghosh 1996, 'Allocation and binding in data path synthesis using a genetic algorithm approach', 9th International Conference on VLSI Design, January 3–6.

Mascull, Bill 1964, *Key Words in Business,* Collins Cobuild, 1996, McGraw-Hill, New York.

McChesney, R. 1997, *Telecommunications, Mass Media and Democracy*, Oxford University Press, USA.

McLuhan, M. 1984, *Understanding Media: The Extensions of Man*, McGraw-Hill, New York.

McLuhan, Marshall 1962, The Gutenberg Galaxy: The making of typographic man, Routledge, London.

McLuhan, Marshall1964, Understanding Media: The extensions of man, Mentor, New York.

Michaelson, H.B. 1990, *How to write and publish engineering papers and reports*, 3rd edn, Oryx Press, Phoenix.

Michell, J.A. 1968, *Writing for Professional and Tehchnical Journals*, John Wiley and Sons Inc., New York.

Mulligan, Geoff 1999, *Removing the Spam: E-mail Processing and Filtering*, Addison-Wesley, Longman.

Nye, D. 1995, *The Technological Sublime*, MIT Press, Cambridge, MA.

O'Connor M. and Woodford F.P. 1975, *Writing Scientific Papers in English*, Amsterdam; New York: Associated Scientific Publishers.

O'Connor M.1991, *Writing Successfully in Science*, Harper Collins Academic, Hampshire, England.

Princeton Language Institute and Hollanders, Joseph, Patrick Hanks, and Jim Corbett 1986, *Business Listening Talks*, Cambridge University Press.

Rathbone, Robert R. 1985, *Communicating Technical Information: A New Guide to Current Uses and Abuses in Scientific and Engineering Writing,*, 2nd edn, Addison-Wesley, Reading, MA.

Ravikumar, C.P. and G. Hetherington 2004, 'A Holistic parallel and hierarchical approach towards design-for-test', International Test conference.

Roland C.G. 1971, *Good Scientific Writing; an Anthology*, American Medical Association, Chicago.

Rubens, Philip (ed.) 2005, *Science & Technical Writing: A Manual of Style*, 2nd edn, Sage Publications, Routledge, New York, London.

Shakespeare 1914, *Hamlet*, Act I, Scene III, In 85, The Oxford Shakespeare, Oxford University Press.

Shakespeare 2005, *Julius Caesar*, Act III, Scene II, ln. 12–260, Oxford University Press, India.

Shaw, Harry 1993, *Errors in English and Ways To Correct Them,* 4th edn, Harper Perennial, New York.

Sherzer, Margaret D. 1986, *The Elements of Grammar,* Collier Books, New York.

Sina, R.P. 2002, *Current English Grammar with Composition*, Oxford University Press, New Delhi.

Sood, Madan 2004, *Précis Writing,* Goodwill Publishing House, New Delhi.

Strunk, W. Jr, and E.B. White 1979, *The Elements of Style*, 3rd edn, Macmillan, New York.

Teeter, D .L. et al. 1989, *Law of Mass Communication*, 6th edn, New York Foundation Press, Westbury.

Trelease, S. F. 1958, *How to write scientific and technical papers*, Williams and Wilkin Co., Baltimore.

Tunstall, Joan 1999, *Better, Faster E-mail: Getting the Most Out of Email,* Allen and Unwin.

Tunstall, Joan 1999, *Easy E-mail,* Allen and Unwin.

Turner, Stuart 1986, *Thorsons Guide to Public Speaking*, Thorsons Publishing Group, New York.

Weiss, E. H. 1982, *The writing system for engineers and scientists*, Prentice Hall, Englewood Cliffs, New Jersey.

Weissman, Jerry 2003, *Presenting to Win, The Art of Telling your Story*, Prentice Hall, New Jersey.

Wilkinson A. M. 1991, The *Scientist's Handbook for Writing Papers and Dissertations*, Prentice Hall, Englewood Cliffs, New Jersey.

Williams J. M. Style 1990, *Toward Clarity and Grace*, University of Chicago Press, Chicago,

Woodford F.P. 1986, *Scientific Writing for Graduate Students: A Manual on the Teaching of Scientific Writing. Bethesda*, Council of Biology Editors, MD.

Wren, C.L. 2003, *The English Language*, Vikas Publishing House Pvt. Ltd, New Delhi,

Zelazny, Gene 2001, *Say It With Charts*, 4th edn, McGraw-Hill, New York.

Zinsser, William 1994, *On Writing Well: An Informal Guide to Writing Nonfiction*, 5th edn, Harper Perennial, New York.

Index